BEING AN EFFECTIVE CONSTRUCTION CLIENT

Working on Commercial and Public Projects

CW01497232

RIBA ₪ **Publishing**   Edited by Peter Ullathorne

Being an Effective Construction Client

© RIBA Enterprises except Chapter 3 © Joanna Eley, Chapter 7 © Rab Bennetts and Chapter 11 Perspective © Dr Julian Critchlow

Published by RIBA Publishing, part of RIBA Enterprises Ltd, The Old Post Office, St Nicholas Street, Newcastle upon Tyne, NE1 1RH

ISBN 978 1 85946 576 9

Stock Code: 83981

The right of RIBA Enterprises Ltd to be identified as the Author of this Work has been asserted in accordance with the Copyright, Designs and Patents Act 1988 sections 77 and 78.

British Library Cataloguing-in-Publication Data

A catalogue record for this book is available from the British Library.

Publisher: Steven Cross
Commissioning Editor: Sarah Busby
Production: Richard Blackburn
Designed, Typeset and Illustrated by HD-design
Cover Design: Tom Cabot/Ketchup
Cover image: Getty Images News
Printed and bound by W&G Baird Ltd in Great Britain

While every effort has been made to check the accuracy and quality of the information given in this publication, neither the Author nor the Publisher accept any responsibility for the subsequent use of this information, for any errors or omissions that it may contain, or for any misunderstandings arising from it.

RIBA Publishing is part of RIBA Enterprises Ltd.

www.ribaenterprises.com

Contents

Dedication

This book is dedicated to all those clients past, present and future who have achieved or have yet to realise their ambitions and are equipped with the necessary vision, skill and courage to create new environments.

So is civilisation sustained.

Acknowledgements

I acknowledge with grateful thanks the work and patience of Sarah Busby, the Commissioning Editor formerly at RIBA Publishing. Similarly, I wish to thank her colleagues who have also worked long and hard to produce this book.

I have thankfully been able to draw on the resources and publications of the RIBA and its senior staff.

I am indebted to Richard Saxon and Gren Tipper for their advice and mentoring during the creation of this book.

The scope of this publication allows an introduction to the world of 'clientship', and other volumes and sources provide a greater level of specialist detail should this be required.

Finally I have to thank all those listed in detail in the section 'About the editor and contributors', without whom this book would have remained an idea.

Peter Ullathorne
London
September 2015

About the editor and contributors

The editor and chapter authors

■ Peter Ullathorne

Peter Ullathorne is a chartered architect and an RIBA Accredited Client Adviser at The Ullathorne Consultancy. He started with a Meccano set and never looked back. Since 1974 he has worked with a number of leading firms including Piano + Rogers, SOM, YRM, DEGW, as a partner at Gensler and HOK on major public and private sector projects and at Navigant Consulting. He led the largest PFI accommodation project for a UK government department. He taught at the University of Cincinnati. In 1985 he co-founded First Architecture plc, the first public company of architects. He established his practice, The Ullathorne Consultancy Ltd, in 2009, providing architecture, client advice and expert investigation services for dispute resolution. He was a ministerial adviser to the Rt Hon David Miliband MP. Additionally, he is a magistrate, an examiner in Professional Practice at the Architectural Association School of Architecture in London and is a member of the RIBA's Professional Conduct Panel. He is a Freeman of the City of London and has an entry in Debrett's *People of Today*.

■ Richard Saxon CBE

Richard Saxon is an RIBA Accredited Client Adviser. In 2005 he was appointed Principal at the Consultancy for the Built Environment and is an authority on the strategic value of BIM. He wrote the Construction Industry Council publication *Growth through BIM* for the Department for Business, Innovation & Skills in 2013 and the forthcoming *BIM for Construction Clients* for RIBA Publications. He is Chairman of the Joint Contracts Tribunal and a director of BLP Insurance. Richard was formerly Chairman of the international design consultancy BDP, Vice-President of RIBA and Chairman of Be, Collaborating for the Built Environment now part of Constructing Excellence. He was awarded a CBE in 2001 for services to architecture and construction.

■ Gren Tipper

Gren Tipper is the CEO and Director of the Construction Client's Group, where he works for leading clients in the industry, promoting best practice and working with the industry to improve efficiency and conditions for all involved in the construction process. Gren is an active member of the Construction Leadership Council Delivery Group and works with the Strategic Forum for Construction, developing and implementing the industries' strategy for construction – 'Construction 2025'. He is a non-executive director of the Construction Skills Certification Scheme, Constructing Better Health, the Considerate Constructors Scheme and is a member of the Chartered Institute of Building.

Gren has always been passionate about the construction industry, supporting opportunities that enable clients to clients can make a difference to outcomes by developing better ways of working.

■ Susie Gray

Susie Gray is a professionally qualified surveying graduate with over 25 years' experience, including in-house corporate real estate, agency and landlord roles, providing an all-round perspective from both the customer and supplier viewpoint. Susie is a director of Empress Consulting, specialising in advising clients on corporate real estate strategy and efficient use of space. Susie is a visiting lecturer at Henley Business School.

◼ Joanna Eley

Joanna Eley is a director of AMA and an architect with over 30 years' experience of developing strategies for space and facilities management. She is a trustee for the Usable Buildings Trust, a facilitator for the HEDQF post-occupancy processes and a RIBA Accredited Client Adviser. She graduated in PPE at Oxford, then architecture at the University of Pennsylvania and the University of London. She has worked with Davis Langdon Consultancy, User Research, Useable Buildings, William Bordass Associates and DEGW. Joanna specialises in buildings for education, offices, museums and housing. Her clients include large public and private sector organisations, many higher education institutions as well as smaller groups and charitable foundations. She is co-author of *Understanding Offices: What Every Manager Needs to Know About Office Buildings,* published by Penguin Books in 1995, and *Office Space Planning: Designing for Tomorrow's Workplace,* published by McGraw-Hill in 2000, as well as writing for the technical press.

◼ Ben Hughes

Ben Hughes is the founder and director of Capital Projects Consulting Ltd. His chapter for this book was written while he was an Equity Partner of EC Harris' specialist Contract Solutions business. He is a chartered member of the Royal Institution of Chartered Surveyors (RICS), Chartered Institute of Procurement & Supply (CIPS) and has an MBA. He has led and delivered procurement and commercial management services for multi-billion pound programmes and projects, and has headed procurement and contract management teams for other top consultancies, the Ministry of Justice and Network Rail, including the management of tender portfolios of £5bn value and the post-contract management of works, services and goods of £600m per annum.

◼ Adrian Dobson

Adrian Dobson is Director of Practice at the RIBA. He is a chartered architect with practice experience primarily in the education and community sectors. He has also taught in higher education and carried out research in building information modelling. Adrian has been closely involved in the development of the RIBA Plan of Work 2013. He is the author of *21 Things You Won't Learn in Architecture School,* published by RIBA Publishing in 2014.

◼ Dale Sinclair

Dale Sinclair is AECOM's Director of Technical Practice, Architecture, responsible for work in EMEA (Europe, Middle East and Africa) and India. Dale's core expertise is the delivery of large-scale projects. He has designed multi-disciplinary processes that successfully deliver projects on time, budget and in line with the client's quality requirements. He passionately believes that the lead designer's role is central to this goal.

He is the RIBA Vice President responsible for Practice and Profession, and an RIBA Trustee and Councillor. He was the chair of the RIBA Large Practice Group from 2007-13 and the editor of the BIM Overlay to the RIBA Outline Plan of Work 2007. As well as editing the RIBA Plan of Work 2013, Dale chaired the RIBA task group responsible for developing it, assisted in the development of the online tools and authored the supporting publications, including the guide to using the RIBA Plan of Work 2013 and *Assembling a Collaborative Project Team* (2013). His publication in 2011 for RIBA Enterprises: *Leading the Team: An Architects Guide to Design Management,* is directed at those who share the objectives of delivering projects more effectively. He is a member of the Construction Industry Council BIM Forum and on the board of BuildingSMART UK. He regularly speaks on the RIBA Plan of Work 2013, BIM and on the future of the built environment industry.

▣ Rab Bennetts OBE

Rab Bennetts co-founded Bennetts Associates in 1987 with his partner Denise Bennetts and provides overall design direction to the firm. Recent projects include the Royal Shakespeare Theatre in Stratford-upon-Avon, the Jubilee Library in Brighton and London Borough of Camden's new headquarters at King's Cross. Rab is also involved in research projects, professional committees and construction education outside the firm and is a board member of the UK Green Building Council, a trustee of the Design Council and a director at Sadler's Wells Theatre. He was awarded an OBE for services to architecture in 2003.

▣ Tom Taylor

Tom Taylor is a highly experienced, qualified and respected project manager, and is a joint founder of Buro Four, Principal of Dashdot and the current President of the Association for Project Management (APM). He is an author and broadcaster. He is a recipient of the 2009 APM Sir Monty Finniston Award and *Project* magazine's Innovative Project Manager of the Year Award 2007.

▣ Gary Wingrove

Gary Wingrove is a chartered surveyor and is Projects and Construction Director for BTFS (BT Facility Services) where he oversees the delivery of all projects; from day-to-day small works to major capital construction projects across BT's 80m square foot UK portfolio of nearly 8,000 buildings. Before joining BT he worked client-side for Morgan Stanley and UBS. He is responsible for managing the construction of new facilities as well as rationalising and refurbishing the existing BT portfolio, delivering efficiencies and cost savings across both the commercial office and operational estates. He manages BT's relationships with outsourced service partners on capital expenditure projects. In 2012 he was elected President of the British Council for Offices (BCO) and was Chairman of the 2011 BCO Conference held in Geneva. In 2013 he was made a Fellow of the Royal Institution of Chartered Surveyors.

▣ Murray Armes

Murray Armes is a chartered architect and founder of Sense Studio, with over 30 years' experience in the construction industry. Murray is also a chartered arbitrator, adjudicator, mediator, dispute board member, FIDIC President's List International Adjudicator and expert witness. He has worked on projects and cases around the world and in the UK, having been instructed on over 100 disputes and given evidence in numerous cases. He advises clients and is involved in conferences and training in dispute avoidance. He has contributed widely as author to many professional works and delivers seminars internationally.

Janet Young

Janet Young is a chartered building surveyor with 25 years' experience in the public and not-for-profit realms. She has worked in the education, housing and commercial sectors and has particular experience of secure buildings through her work with the Ministry of Justice and the Foreign & Commonwealth Office. Janet is now Director of Estates at the Ministry of Justice and is the Ministry's Head of Profession for Project Delivery. She leads multi-disciplinary teams who look after one of the government's largest estates comprising courts, prisons and probation buildings. Janet has served on the British Property Federation's Construction Committee, the Building Regulations Advisory Committee and has been an external examiner for the MSc and BSc building surveying courses at Leeds Metropolitan University.

Ruth Reed

Ruth Reed is a past president of the RIBA (2009–11) and a chartered architect with a Master's degree in landscape. She is a director of Green Planning Studio with responsibility for architectural services, built environment and landscape issues. She has been Vice-President of Membership of the RIBA and President of the Royal Society of Architects in Wales. She is a judge for architectural awards including the Stirling Prize. Ruth is a Vice-chairman of the Construction Industry Council Board and Chair of the RIBA Planning Group.

She has acted as a professional witness in planning appeals since 2007 on behalf of her firm. She is a former Professor of Architectural Practice and Director of Professional Studies at the Birmingham School of Architecture, Birmingham City University.

Ruth authored the *Town Planning: RIBA Plan of Work 2013 Guide,* which was published in November 2014.

About the Perspective authors

Professor Sir David Omand GCB

Professor Sir David Omand is a University of Cambridge graduate in economics, has an honorary doctorate from the University of Birmingham and has recently completed a degree in mathematics and theoretical physics with the Open University. He is a member of the editorial board of *Intelligence and National Security* magazine. With Dr Michael Goodman of the Department of War Studies at King's college, London, he is responsible for delivering training to government intelligence analysts and he lectures regularly to BA and MA level classes in Intelligence Studies.

In 2002 Sir David Omand was appointed the first UK Security and Intelligence Coordinator, responsible to the then prime minister, Tony Blair, for the professional health of the intelligence community, national counter-terrorism strategy and homeland security. He served for seven years on the Joint Intelligence Committee. He was Permanent Secretary of the Home Office from 1997 to 2000, and before that Director of GCHQ (the UK Sigint Agency). Previously, he served in the Ministry of Defence as Deputy Under Secretary of State for Policy, and was Principal Private Secretary to the Defence Secretary during the Falklands conflict. He served for three years in NATO in Brussels as the UK Defence Counsellor. He has been a visiting professor in the Department of War Studies since 2005–06. He is the Senior Independent Director of Babcock International Group plc.

■ Mark Hackett

Mark Hackett is a chartered quantity surveyor specialising in dispute resolution. He has been involved in a wide variety of construction schemes and has given evidence in numerous UK trials and international arbitrations. Mark is also the co-author of *The Presentation and Settlement of Contractors' Claims,* published by Routledge in 2000, and the co-editor of *The Aqua Group Guide to Procurement, Tendering & Contract Administration*, published by Wiley-Blackwell in 2007.

■ Professor Hans Haenlein MBE

Professor Hans Haenlein is a leading expert in the planning of education and community facilities. He has over 40 years' experience as an academic and as a practicing architect. From 1976–91 he was Head of the School of Architecture at South Bank University and later Dean of the Faculty of the Built Environment. In 1987 he was awarded an MBE for services to architecture. Since 1991 he has been Professor of Architecture at the University of Reading where he has designed a concept building for the International Centre for Inclusive Environments and carried out research into various aspects of briefing, design management, construction management and inclusive environments. He has been responsible for the design of many schools, community buildings and urban regeneration projects. Since 1985 he has been President of the Hammersmith Society and is a RIBA Accredited Client Adviser. In 2013 he was elected President of the First Education Alliance.

■ Gareth Hird

Gareth Hird is a chartered surveyor and is responsible for McBains Cooper's commercial activities. He is an alumni of the Kellogg School of Management, a Liveryman of the Company of Chartered Surveyors and a Freeman of the City of London. McBains Cooper was founded in 1785 and provides a wide range of services to the public, private and institutional realms, worldwide and across all significant sectors. Gareth provides specialist expertise, including due diligence advice to funders and investors, and has advised some of the world's most respected financial institutions and funds.

■ Michael Darner

Michael gained a Masters in Architecture at the University of Denver. He also began work as an office boy at Gensler and Associates where he has spent his entire career, becoming a partner in 1993. Starting with a reconstruction project of a tower modelled after the Piazza San Marco in Venice (except this one was in Denver), his career has included leading the design, management and construction of a wide variety of projects, including the Beverly Hills Country Club, Sony Pictures Studios renovation in Culver City, Sony Lincoln Square Theater in New York City, Studio Babelsberg renovation in Berlin, Liverpool Airport's renovation, Leavesden Studios' creation in Hertfordshire, Mashreq Bank Headquarters in Dubai, the Gulf Investment Bank's Headquarters in Kuwait, the American Film Institute Silver Theater in Washington DC, and currently the 34-storey Landmark Tower Apartments in Los Angeles. Michael says, 'I've had wonderful client relationships over the years, which I think springs from a genuine interest in the people

I've met and the problems they've asked me to solve. Creating for that purpose has been its own very satisfying reward.'

◼ Dr Timothy Stone CBE

Dr Timothy Stone is a visiting professor in University College London's International Energy Policy Institute and is also the senior expert non-executive member on the board of the European Investment Bank, a non-executive director of the Anglian Water Group and a non-executive director of Horizon Nuclear Power. He is also a member of a number of advisory boards for infrastructure investors and funds. Until recently, Tim was the Expert Chair of the Office for Nuclear Development in the Department of Energy and Climate Change and the Senior Adviser to successive secretaries of state responsible for energy. He served five different secretaries of state in two different governments. His previous career was as the Chairman and founder of KPMG's Global Infrastructure and Projects Group, a managing director of SG Warburg & Co in New York and London, and a managing director of Chase Manhattan Bank in New York. He was appointed a Commander of the British Empire in the 2010 Birthday Honours List for services to the energy industry.

◼ Peter Stewart

Peter Stewart is a chartered architect and the Principal of Peter Stewart Consultancy, which he founded in 2005, providing expert advice in the fields of architecture, urban design, planning, townscape and the historic environment. During 15 years in practice he was responsible for large office and residential projects from inception to construction. In 1997 he became Deputy Secretary of the Royal Fine Art Commission (RFAC) and in 1999 joined CABE, now Design Council CABE. Until 2005 he was Director of CABE's design review programme. His work at the RFAC and CABE involved him in advising on many of the most significant projects in the country.

He was the principal author of a number of CABE publications, including *Design Review*. He drafted the original CABE/EH *Guidance on Tall Buildings*. Peter has served as an expert witness at many planning inquiries and is a member of the London Legacy Development Corporation Quality Review Panel. He has served as Chair of the regional design review panel for the East Midlands, as a member of the London Advisory Committee of English Heritage and as Chair of the Planning Group of the RIBA.

◼ Anthony Slumbers

Antony Slumbers is a specialist in the use of internet technology in the real estate sector; his particular skills are in designing online applications that encourage collaboration and efficient business processes. He founded Estates Today in 1995. He was a founding equity partner with Broadgate Estates of Vicinitee.com, the community and property management software with which some two million people a year interact. In 2007 he founded Glasnost21, the leading collaboration financial software SaaS. Glasnost provides project and image management as well as CRM and email marketing tools. He has also been a founder partner in CityOffices.net since 1999. Today, Antony's primary focus is on designing and developing mobile and cloud based applications, with a particular emphasis on collaboration, localised data and social networks.

■ Roger Camrass

Roger Camrass has enjoyed a 40-year career in change management and digital technologies. After graduating from the University of Cambridge, he was a post-graduate research Fellow at MIT in the 1970s as a pioneer of today's internet. Roger has devoted his life to helping companies large and small to exploit emerging digital technologies, including cellular communications, cloud computing and the Internet of Things. His career appointments include partnerships with leading consulting groups such as EY and Arthur D Little, management roles with the Stanford Research Institute and Fujitsu, and startups such as Butler Cox and Mozaic Services. He is author of the book *Atomic: reforming the business landscape into the new structures of tomorrow,* published by Capstone Publishing in 2003, and a visiting professor at the University of Surrey and ESADE Business School.

■ Ken Anderson

Ken Anderson is a senior partner at the investment bank Greenhill & Company, joining the team in 2014 after seven years with UBS in London. At UBS, Ken was first a managing director and then a vice-chairman. Ken works with clients in a variety of industries with an emphasis on healthcare. Before joining UBS he was the first Commercial Director General at the Department of Health and a member of the department's management board. In this role, he was responsible for the introduction of non-NHS clinical services, negotiating the PPRS (Pharmaceutical Price Regulation Scheme), the UK's medicine purchasing agreement worth £12bn annually and rationalising the £14bn NHS supply chain including the £20bn outsourcing of NHS Logistics. Ken has over 16 years' experience within healthcare operations in the USA, UK and continental Europe, and is Adjunct Professor of Finance at the Imperial College Business School.

■ Dr Julian Critchlow

Dr Julian Critchlow is Head of Construction and Energy at solicitors Payne Hicks Beach and specialises in construction law. He acts for developers (both international and private), main contractors, subcontractors and consultants in respect of both project advice and disputes. Julian is an arbitrator in both domestic and international disputes, frequently appointed in Department of Immigration and Citizenship arbitrations. His PhD thesis from King's College London concerned arbitration. He is a Fellow of the Chartered Institute of Arbitrators, a TeCSA registered adjudicator and a CEDR Accredited Mediator.

Julian's numerous publications include *Arbitration Forms and Precedents* (jointly with Professor Rob Merkin) and *Making Partnering Work in the Construction Industry.* He is also a well-known speaker on construction issues. A number of his cases have been reported. Julian has been included in 'Chambers UK' and 'Legal 500' for many years, Chambers most recently stating that his clients are impressed by his 'extensive knowledge, sound judgment, clear explanation and astute tactics'.

■ DDJ Stuart Kennedy

Stuart Kennedy has a First Class Honours degree in law, is a practising barrister in England and Wales, a barrister and solicitor in St Vincent and the Grenadines, and is a Deputy District Judge (DDJ) on the South East Circuit in England. He is a chartered arbitrator and is on the Panel of Arbitrators, Adjudicators and Mediators for the Chartered Institute of Arbitrators and the Royal Institution of Chartered Surveyors. He is a Fellow of the Royal Institution of Chartered Surveyors, a Fellow of the Chartered Institute of Arbitrators, and a member of TecBar, the Society of Construction Law and of the Western Circuit.

Stuart's practice as a barrister focused on construction and he is ranked as a leading junior in Legal 500 2014 and by *Chambers and Partners Legal Directory 2015*. He has been involved in a number of major international projects, including the new airports in Hong Kong, Oslo and Athens, as well as projects in Europe and the Middle East, including the new Wembley Stadium in London. He appeared for the claimant in a large arbitration in Hong Kong in a claim against the government. In addition to sitting as a DDJ, Stuart now acts as mediator, adjudicator or arbitrator in disputes in the UK, the Caribbean and internationally. Stuart has been appointed to undertake Early Neutral Evaluations on major construction disputes including a £70m dispute on a hospital project and a £15m dispute on a city centre redevelopment. He has appeared in a number of reported cases involving challenges to arbitrator's and adjudicator's awards. He sits as a mediator on various types of commercial disputes with particular emphasis on property, construction, engineering and energy disputes and also lectures at conferences and seminars and has run training courses on mediation and arbitration in London, Kenya and the Caribbean.

◼ Ty Robinson

Ty Robinson has worked within the infrastructure and support services sectors for more than three decades and has held senior posts in both public and private sectors. Ty provides support to private and public sector clients looking to transact in the space where these two sectors intersect. He spent 20 years in construction where he helped deliver major infrastructure projects, including GCHQ and GlaxoSmithKline's United Kingdom research headquarters. Ty served as Chief Operating Officer of the United Kingdom's Department of Health Commercial Directorate and, subsequently, with Partnerships UK provided advice and guidance to local authorities and the Homes and Communities Authority on a programme of national regeneration projects. In the private sector Ty was Managing Director of the healthcare division of Navigant Consulting for Europe. He has significant sector experience in defence, security, housing and urban regeneration, health, custodial and transport sectors, successfully delivering projects and large-scale programmes involving public/private sector collaboration.

◼ Paull Robathan

Paull Robathan is a technology strategist and designer. He was responsible for the furniture, interior design and technology integration of complex financial trading environments all over the world. He spends much of his time now working in the field of social housing, including doctoral research at University of the West of England and working with the Schumacher Institute.

Foreword

by Richard Saxon CBE

These are challenging times to be a client for a building. Pressures are rising, ideas and attitudes are changing and technology is offering new approaches. Few clients are expert in their role, hence the use of the word 'client', meaning a dependant. Those that are expert build constantly and confidently, often in infrastructure businesses, commercial developers or institutional bodies like universities. Even these continuous clients compare notes with peers. The majority of clients are less experienced and it is to these that this book is addressed, though all may find value in it.

The world is in the throes of a building boom unlike any before. Billions of people are moving into cities over the next few decades and urbanisation is exploding across the developing world. This is putting global pressure on resources for construction, but also creating new sources of low cost components. The challenge of sustainability is growing, both from regulation and from corporate reputation factors. In the UK the capacity of the ndustry, to design, manufacture and assemble buildings has been greatly reduced by the long recession between 2008 and 2012. Returning demand is hitting supply constraints.

Over the 20 years since the Latham Review of the Construction Industry, clients have moved towards collaborative forms of working with designers and contractors, only to abandon them for fixed-price tenders when the cut-throat market of 2008-12 seemed to offer better deals. Now the sellers have market power again and there are deeper reasons for collaboration: the government has intervened to move the market. Its 2011 Construction Policy sought to drive out cost and increase value from construction for central government clients. It introduced tools of great use to all clients: benchmarking to reveal the achievable cost levels for building types, 'Soft Landings' to ensure that buildings are delivered in proper working order and that their reasons for construction are proved to be achieved, and Building Information Modelling to support a much more effective design and construction process, together with the basis for good whole-life management of the facility. Whole-life performance has become definable and deliverable with the new tools available.

This central initiative was pushed further in 2013 by the creation of an Industry Strategy for Construction, agreed by the government and leaders of the industry in partnership. It sets goals for 2025 which include reducing the capital and whole-life costs of buildings by 33%, decreasing carbon emissions by 50% and also halving the time taken to move from decision to build to completion. These apparently 'stretching' targets are actually almost within the benchmarked range observed between projects with expert clients and those without.

Client skills are the key to success, yet they don't teach these skills in business school. Less experienced clients can emulate the expert ones if they follow the precepts in this book. The leading requirement is to prepare properly. Time spent on reconnaissance is seldom wasted, as the Duke of Wellington opined. Is there a clear business case for a building or could the need be met another way? Who are the internal and external stakeholders and are they participating properly? Is the proposed site or building for conversion feasible? Are the resources needed available, both to build and to operate the facility? What will success look like? Has the client got the time and capacity to play the client role properly? How much risk are they able to manage?

Clients need not be alone. Advisers are necessary and can fill out your skill set. Executive horsepower can be hired. Support groups for clients exist. British Standards provide sound workbooks. The new RIBA Plan of Work provides a framework for all the steps required. But the final decisions are the client's. They need to understand what the 'value proposition' of their project is: how it delivers benefits that suitably exceed costs. Value has many dimensions: use, exchange, image, environmental, social and cultural. It is embodied in the design they select, not in the work on-site. They need to be able to explain the value proposition to internal and external participants and to recognise when they are hitting or missing their goals. Armed with the insights in this book, clients should be more confident of success.

Foreword

by Gren Tipper
Director at Construction
Clients' Group,
Constructing Excellence

Clients come in all shapes and sizes. Some engage in construction on a regular basis and as a result their organisational structure and selected business partners reflect this. There are, however, many clients that occasionally need to engage the industry to satisfy a business or infrastructure need. This book is therefore structured to assist all clients. It is written by experts in their field and is injected with real life experiences from clients who have witnessed both excellent and, on occasions, poor outcomes from the construction process.

This book assists clients in understanding how the construction industry works, what they need to do at each stage of the construction process and, importantly, where to find help in what can seem like a minefield to those who have little understanding of procuring construction, other than experiences gained in a more domestic setting – which it must be said is not typical of the professional industry that delivers the majority of our built environment. Having said this, those who consider themselves to be experienced clients can learn from others; best practice is ever-evolving rather than static. One size certainly does not fit all when it comes to engaging in construction activities, but by applying best practice in a measured manner and adapting to suit a particular need, it can be really rewarding with little effort and with avoiding what can be painful and costly consequences when the wrong decisions are made.

This is a must-read guide for clients who want to realise the best possible value from their commissions. Engaging the right people with the right expertise at the right time is key to a successful project, as is understanding the risks to be managed and how best to manage and mitigate these risks by selecting the most appropriate procurement route. Clients, and those engaged in delivering construction, all share the desire to run successful and profitable businesses. Understanding this and ensuring that all are treated fairly is the foundation for building the right relationships.

History tells us that the best construction solutions come from an environment where the client places itself at the centre of an integrated team that works collaboratively. Clients are required to demonstrate leadership from the outset, being clear on what they require and providing the means to measure success and to reward excellent performance. Working as a team requires nurturing. It does not happen unless those leading demonstrate a culture of openness and trust which will only gain traction if the client makes a clear statement on these values at the outset and assumes the primary leadership position. Being a client is not a spectator sport, cheering when things go well and booing when not. Client leadership is covered extensively within this book and for good reason; it makes a real difference when done properly.

The effort required by the client is proportional to the complexity of the proposed project and provided this is understood and the right expertise is brought together, then the construction process can be an enjoyable experience for both frequent and occasional clients, leading to an asset the end users will be delighted to own and use as part of their businesses.

It was no accident Sir Michael Latham chose the name 'Constructing the Team' for his report that proposed reforms to the construction industry back in 1994. Twenty years on, we have a catalogue of successful projects in Britain, founded on partnering in a collaborative environment that makes us the envy of the world. We just need to spread this best practice far and wide, and fully take advantage of the benefits it offers.

Chapter 1 Introduction

by Peter Ullathorne

"Those who wish to build in the United Kingdom must be effective in a mainly negative environment where public opposition, local government inertia and comprehensive legislation must often be overcome or complied with in order to achieve projects."

Clients and buildings

Buildings, buildings and yet more buildings. Distinct from the landscape, buildings are among the numinous structures that enable our civilisation. Clients seek to organise space so that we can live, work, repair, defend and enjoy ourselves. They are pre-eminent – having the basic initiative and constant drive to achieve buildings. They assemble resources and consents to create environments and are the apex of the construction pyramid and yet very little light has been cast on their activities, the skills and the techniques they need to use. Mainly, the clients are the quiet and understated magicians who achieve so much.

The target readership

This book will inform and guide clients in the commercial, public and institutional sectors. While there are countless books on architects, design, engineers and buildings, there are very few that focus on what could be called 'clientship' – in other words, all that it takes to be an effective client. This book provides essential information to clients who are new to, or have limited experience in, clientship, as well as the very many who serve and deal with clients, including students of the professions and those who have ambitions to be clients. A distinguished faculty of authors and commentators who are either clients, or who advise clients, have worked together on this book. Our authors provide essential insight, guidance and information for those taking responsibility for building projects in both the public and private sectors.

Our Perspective authors provide thoughtful - and sometimes challenging opinions on the subject of clientship.

The process of producing buildings has become highly complex and exacting, and clients need a wide range of consultants, advisers and constructors to achieve success. For clients to be effective – and this key criterion is expressed in the title of this book – they need to be served by those who are experienced, creative and skilled, who are right for the particular assignment and who know what support their clients really need. There are no university courses in clientship and with temerity this book aims to start to fill that lacuna.

Attitudes for change

Those who wish to build in the United Kingdom must be effective in a mainly negative environment where public opposition, local government inertia and comprehensive legislation must often be overcome or complied with in order to achieve projects. The developer Qatari Diar felt the full force of majestic opposition when it attempted to gain consent to redevelop the Chelsea Barracks site, which is only now under way after years of design, redesign and negotiations. Major projects are notoriously difficult to progress in the UK as politicians are extremely wary of supporting these against even a whisper of public dissent, and prospective clients for projects large and small should be aware of the disproportionate influence of pressure groups. Clients will always meet resistance and it is reasonable to propose that they must have political skills as a basic requirement in order to move projects through the resistance of inevitable opposition. Three projects are evidence of this: HS2, an additional runway for a London airport and Crossrail 2, all infrastructure projects that we have needed for years and are mired. It is in this climate of opposition to development and change that clients must press their cases to achieve projects – to be effective.

Architects and clients

The RIBA Client Liaison group, led by Nigel Ostime, was established in 2014 by RIBA President Stephen Hodder to understand the clients' perception of architects and the value they bring to the project team. This builds on the work completed some 20 years ago by the then President of the RIBA, Dr. Frank Duffy, who commissioned a landmark strategic study of the profession. So the profession is able to keep abreast of the real needs of clients in key sectors. RIBA Accredited Client Advisers are actively working with clients from the earliest stages of their projects (Stage 0 onwards) and provide additional and valuable feedback to the profession. The RIBA document 'Leading architecture: The RIBA's Strategy 2012-2016' (downloadable from www.architecture.com) provides a list of strategies and goals for the RIBA and for architecture related to clients whose objectives are still midway on the waves.

Authors and topics

In this book, clients are given the vital message that they must define the outcomes they want in order to discover the best ways of achieving them. The first reactive response to a management problem may be to build. However, a more considered response might be to do something very different. Susie Gray explores topics of feasibility, business case and funding in detail. If a building project is the best response, then Joanna Eley describes the briefing process – the essential DNA of any project. Ben Hughes examines the art and science of procurement and describes the principal methods that clients can use to purchase their buildings. Clients will need to assemble a project team that is capable of responding to the challenges and opportunities of the project and Adrian Dobson describes what services are needed and how they can be chosen. Projects need to be organised efficiently and effectively and Dale Sinclair, as the author of the RIBA Plan of Work 2013, describes the reasons, structure and detail of this plan and how it provides a consistent and comprehensive methodology for achieving projects. From a position of authority, Rab Bennetts shines light on the mysteries of the design process. Drawings are the *lingua franca* of communication between professionals, the client and those who construct, and Richard Saxon explains that the construction of a common digital building model, with shared liability for the work, requires productive collaboration. He describes Building Information Modelling (BIM) and gives us a view of the future. Tom Taylor explores the world of project management. From his vantage point as Project and Construction Director for one of Europe's largest companies, Gary Wingrove describes the process of accepting, commissioning and using projects.

The use of formal service and construction contracts practically invite disputes between the parties and expert Murray Armes writes about the laws and clauses that apply to clients, and offers them advice on what they should do if things begin to go wrong.

Janet Young describes the nature of the government client and explains the methods, criteria and processes by which government evaluates and manages projects. RIBA Past President Ruth Reed describes the intricacies of the town planning system.

Perspectives

Our faculty of writers includes a number of distinguished commentators, each providing their unique perspective on clientship. Among them Sir David Omand asks the wise and thoughtful question 'Do you really want us as clients?', while Tim Stone wishes clients would better understand the need for advice rather than process, and value more highly the proactive and creative advice given by independent professionals.

So what makes you a 'client'?

How do you know if you are a client? The accepted definition is that clients are those who are protected by their professional advisers. The relationship between client and professional is usually long term and the degree of trust placed by the client in the work of the professional is significant. In the context of this book, clients have the money and the mission to achieve change by creating environments across a very wide spectrum of scales and types. To be

effective, they have to be astute in business, as well as sustaining and promoting the project vision through all project stages and into occupation. They produce the building and may go on to manage it over time. The players may come and go, but the client remains. Of course, the client is not omnipotent and all-seeing. However, they see the overall picture and ensure that everyone else sees it too. Clients make strategic (and tactical) decisions before and during the gestation period of the project, hold the money and are responsible for the returns on capital employed – and these returns could be as diverse as a decent percentage or enhanced learning outcomes. They are accountable to stakeholders. To be a client is to occupy a position of trust.

Clients need to be aware that almost every building or environment is a prototype and the result of 'purposeful creativity'. Dr Andrea Siodmok, writing in the RSA Journal (Issue 4, 2014) commented that prototyping generates imperfect truths and with the right approach also generates data about the future. Clients and their advisers need to be aware of what it takes to create a prototype that works perfectly first time at full speed without crashing. For those who mistakenly underestimate the importance of design in achieving the right outcomes, the OECD (Organisation for Economic Co-operation and Development) has said that 80% of the impact of any product or service is determined at the design phase. Decisions made in the first 10% of the process heavily influence outcomes. To be effective, clients must value their appointed professionals and constructors for their talents, skills, creativity and gifts.

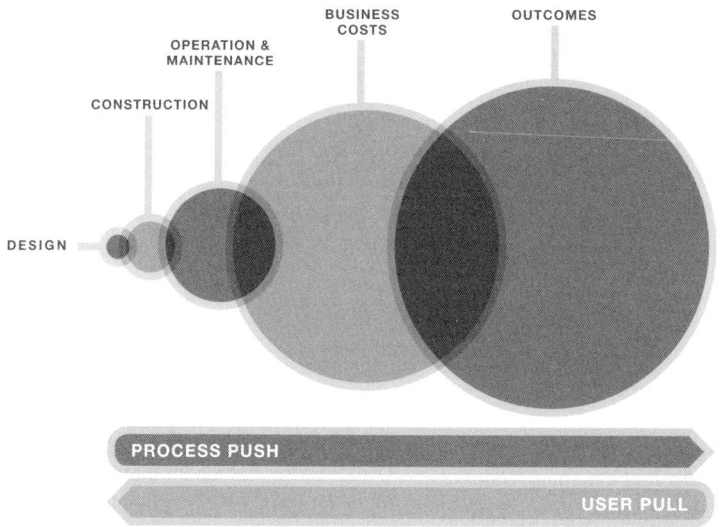

Figure 1.1: The value relationship within the build cycle

Basic leadership skills are needed for the client to be effective and the effective client will transfer leadership qualities to team members. Max De Pree at Herman Miller identified the concept of the roving leader who is there when needed, stepping in and out of the scene as required. Sometimes the client will recognise that some functions can be done better by others and will move aside to allow this to happen. They will recognise when things are going wrong (including team morale) and take swift action.

Clients may wish to transfer as much risk as possible to others. Tony Bingham, writing in the *Estates Gazette* (9 January 2015), criticised the concept of design and construct (D&C) which exposes the client to a system where design is relegated to a backroom function over which the client has little or no control, particularly when the initial designs have been approved and the scheme is taken in for construction drawings and specification to be devised and building done. As he put it, 'designers design, builders build'. It is all too easy for an inexperienced client to be seduced by D&C. However, if the effective client employs competent advisers, the best arrangements can move forward successfully.

Clients who choose architects and schemes though competitions must have the courage and capability to realise the winning scheme, unlike the Cardiff Bay Opera House Trust and the Japanese Government with their Olympic stadium. Clients must be aware of the many dangers to the success of their projects, of losing their client status and of becoming 'customers' of a project over which they have significantly reduced control.

Constructing Excellence

Constructing Excellence provides a vital forum for successful clientship, including achieving value through collaborative work and support for the government's 'Construction 2025' strategy. For example, relationships between the client and supply sides will change significantly and added value will be the basis of payment. The reader is recommended to refer to CE's website at www.constructingexcellence.org.uk for further details.

The above figure shows the fiscal relationship between design, build cost, operating costs, business costs and the financial benefit of the outcomes. Too many clients fall into the trap of minimising the importance of design, denying to them the vastly greater financial advantages arising from good design.

After the crash

Clients need to consider that the entire means of production of the built environment continues to emerge from the worst financial crash since 1929. The fall of Lehman Brothers in 2008 marked the irrevocable and fast-moving collapse of property values, leaving gigantic debt that only bottomed in mid-2009. Almost every type and size of client suffered, with the exception of London's four great family estates (Cadogan, Grosvenor, Howard de Walden and Portman) that actually saw a healthy increase in their values. Major and successful developers suffered gigantic losses. Professionals and the construction industry were hit hard. Public sector work was curtailed.

Clients know that the surviving lending banks have recovered and are lending – now actively looking for investment opportunities. An air of optimism pervades the market. The degree of skill required by clients to raise money at the right price is significant, with developers conscious of changing economies in the Middle and Far East, as well as Russia. Oil revenues are falling, reducing the money supply in some countries. For the public sector, the private finance initiative (PFI) is only attractive when other sources of funding projects are unavailable, generally offering poor long-term value. The supply of public money is restricted at national and local levels. PF2 is a new Treasury model[1] for achieving specific types of public sector projects, offering solutions to the serious

shortcomings of the original PFI concept. Additionally, those bidding for public sector work pay a heavy cost in actual and risk terms, facing heavy competition for any opportunity. Over time this has meant that a significant debt remains in the industry begging to be resolved, which is unlikely with the current (arguably flawed) framework system.

As we distance ourselves from the crash, clients will find it increasingly easier to find senior debt for their projects. Encouragingly, Property Wire reported in January 2015 that there was a significant rise in new lending for commercial property in the UK, particularly in London in the first half of 2014. The construction industry is trying hard to accommodate this growth, which consists of delayed projects and new initiatives. Whilst the government's now defunct BSF[2] programme sustained many professionals and construction firms during the hard years, its replacement is but a shadow and the public estate has shrunk dramatically as Janet Young describes in Chapter 13. Because of shortages of skilled labour and materials against a rising (or pent-up) demand, some prices have increased to the disbelief of the naïve client.

Soft Landings

Clients have a greater chance of achieving a smooth transition from construction to occupation and an optimised performance by using what is called a 'Soft Landings' strategy jointly developed by the Building Services Research and Information Association (BSRIA) and the Usable Buildings Trust (UBT.) For government clients, in May 2011 the

Government Construction Strategy proposed the Government Soft Landings (GSL) initiative to reduce the cost of construction and improved performance for public buildings. Further information on Soft Landings can be found at www.bimtaskgroup.org.

The client and data security

Clients in the commercial and public realms need to protect both their valuable data from competitors and also their organisations from those who wish to either deny their continued service or inflict lasting harm. What is not well understood by many clients is that one form of leak is the unwanted two-way flow of data. Clients involved in creating and using the built environment (both buildings and their interiors) need to be able to assess the risks they face and lead the actions necessary to protect their data, their customers and clients and ultimately their organisations. Cyber assault is a form of warfare waged aggressively by nations and corporations alike.[3] Clients must safeguard their data assets.

The client and physical security

Clients need to be aware at the outset of any project and during their occupation of their real estate of the need for effective physical security. At the time of writing, the threat level in the United Kingdom is at the top end of the 'Severe' category, only a hair's breadth away from the ultimate 'Critical' level. Some clients are more vulnerable to some form of physical attack than others. The deadly assault in Paris in January 2015 at the offices of the magazine *Charlie Hebdo* shows how easily a prepared aggressor can make an attack. Many organisations will, by the nature of what they do, attract some form of aggression.

Clients and research

Effective clients will rely either directly or indirectly through their professionals on the products of research. New forms of contract, working methods (eg RIBA Plan of Work 2013), design technologies (eg BIM), products (eg graphene nanochem) and unconventional computing[4] are examples of directions for the interested client.

A number of organisations continue to address the issue of best practice for clients, including the RIBA, which has published the report 'Client Conversations', providing insights into successful project outcomes. Also the Construction Clients' Group (CCG) within Constructing Excellence has produced six Client Commitment Guides; a code of conduct to advise clients on how to create better value successfully. These contain guidance on client leadership, achieving design quality, commitment to people, procurement and integration, health and safety, and sustainability. Reference to the CCG website is included in the bibliography in Appendix 1. ∎

Notes

[1] HM Treasury: A new approach to public private partnerships (December 2012).

[2] Building Schools for the Future programme, initiated by the Rt Hon David Miliband MP.

[3] The Institution of Engineering and Technology/Centre for the Protection of National Infrastructure 'Resilience and Cyber Security of Technology in the Built Environment' (2013).

[4] Armstrong, R (Professor of Experimental Architecture, University of Newcastle), Super-complexity Control (2013), Riverside Architectural Press/ABC Art Books Canada.

Perspective

by Professor Sir David Omand GCB

Honest dealings

'Are you really sure you want us as clients?' was how, as the Permanent Secretary, I used to probe the senior management of the shortlisted bidders in any major competition. We had by then dreamed the visions conjured up by the Armani-suited black polo neck wearers and in return had given profuse assurances of the importance of the project to ensure sufficient bidders stayed in the competition. Now it got serious. Now would come the period of truth telling on which would rest the success or failure of the project.

Is there mutual understanding of what each party needs to secure from the deal and how it is to be made win-win? Does the bidder appreciate the political environment in which high profile public sector procurement takes place, when the minister can change overnight and the priority given to the project likewise? And that public sector security constraints will add cost and inconvenience? Is there mutual recognition that the cost of time to the public service is very different from the cost to the private sector? Is the client being open about the savings expected from the project, and that the project would not be afforded without them? Is the bidder being honest about what it will take on the part of the client – not least to avoid over-defining solutions to requirements in advance – to release the innovation and energy essential to realise such transformation? And the staff support and priority that will be needed from the client side (almost certainly underestimated)? Even the committed managers will not have recognised how far their thinking is shackled by the way things are done now, nor how they might be transformed by the project. Is it clear that the senior responsible owner will have the authority to make the necessary changes happen?

Once the preferred bidder stage is reached, there is nowhere to hide. Honesty above all else is needed; this breeds confidence that once the contract is signed it will never have to be brought out of the file.

Chapter 2 Project feasibility, business case and funding

By Susie Gray

"It is vitally important that property is aligned with organisational strategy, as property costs are often the second highest cost to an organisation after its staff. Therefore, real estate decisions can have a significant impact on an organisation's ability to deliver its strategy."

Introduction

It is possible to be so entranced by shiny things, be they gadgets, shoes or buildings, that one might be convinced of their suitability without properly considering whether they are chosen for the right reasons. Do they offer the right balance of risk across cost, quality and timing? This book will guide you skilfully through the nuances of sponsoring a building project, but this particular chapter will take you back to the root of your needs to ensure that you can substantiate your decision to build. This is important because commissioning new premises is complex enough to warrant this entire book on the subject and is therefore not a journey to be embarked upon unless you are certain it will deliver a solution that is right for the organisation.

This chapter will set out the five questions you need to ask and answer before making the decision to commission a building project. Questions 1 and 2 relate to understanding the organisation whilst Questions 3 and 4 seek to interpret how those organisational characteristics and needs manifest themselves in property terms. Question 5 is your choice of procurement options.

Question 1: What sort of an organisation is it?

Start by considering the organisation's purpose and strategy. The table to the right shows examples of some common strategic themes and the implications for real estate decisions.

Table 2.1: How strategy aligns with real estate decisions

Organisational and real estate alignment	
Purpose/strategy	**Real estate implications**
Consumer goods company with a sales growth strategy	Growth requires funding and any available capital will be invested in projects that deliver growth. Unless investment into premises can be linked to this growth strategy, it is unlikely to be allocated funds.
Public sector services organisation with a strategy to improve service quality	There are links between service quality and the quality of the environment for those delivering the services, so investment in premises would be aligned here. However, a build project would only be considered if capital funding was available. Otherwise, revenue-funded premises improvement might be a better option.
Branded consumer goods or financial services company with strategy to improve margins	Brand differentiation is a way to increase prices and improve margins. A new build project can support this strategy by providing brand visibility to the public and reinforce the brand value to staff through the quality and design of a bespoke building. Margins could be improved through the provision of more cost-effective space or better support of process flow within the organisation.

9

Organisational and real estate alignment cont.	
Public sector organisation with a strategy to reduce overheads	If there is pressure to reduce annual running costs, then investment in a new building could deliver accommodation that is smaller and cheaper to run. If capital reserves are available, then an owned building will place less strain on revenue funding than a rented one. A move to a cheaper location could also achieve the objective of a lower cost base.
Traditional company seeking to achieve organisational changes	The style, layout and quality of a building can have a significant affect on the culture, effectiveness, cohesion and staff appeal of an organisation. A new environment can be a very effective reinforcement of organisational changes and can help to embed new work practices.

It is vitally important that property is aligned with organisational strategy, as property costs are often the second highest expenditure to an organisation after its staff. Therefore, real estate decisions can have a significant impact on an organisation's ability to deliver its strategy.

As well as alignment with strategy, it is important to understand the implications of real estate on the accounts. Capital expenditure will hit a profit and loss account but the investment may strengthen the balance sheet for years to come. Leasehold commitments are often 'off balance sheet' and the effect of the liability is not noted there. However, the International Financial Reporting Standards are likely to change in 2015 and the payment liability of the whole length of a lease will need to be capitalised. This will have an impact on the balance sheet, which may affect some organisation's ability to raise finance.

Question 2: What is driving the organisational need?

Now, consider how this began. What event triggered the thought that the organisation needed to move or commission a building? It is important to be clear about the reasons so that you can ensure that the chosen solution will meet this need, deliver the desired result and 'keep the vision'. Some common drivers are described here to help you to identify the one that is relevant to the organisation.

Expansion

Although the need for more space is often linked to expansion, it is important to be clear about whether this is genuinely linked to business expansion or just space expansion. In the case of the latter, the impact of a potential rise in overheads needs to be balanced against other improvements to the organisation that can be attributed to the larger property, such

PROPERTY CYCLES

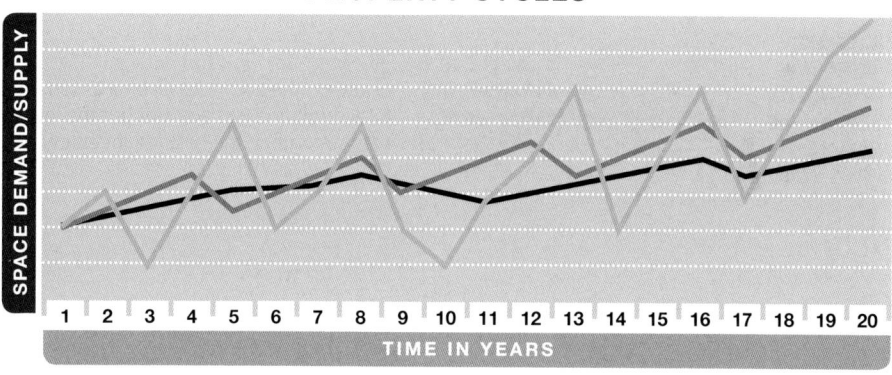

☐ Figure 2.1: Property cycles

━━━━━━━━ PROPERTY
━━━━━━━━ PUBLIC SECTOR
━━━━━━━━ CORPORATE

as improved staff morale, staff acquisition and retention or increased levels of interaction and innovation. If the increased costs cannot be justified, then consider whether more could be achieved from the existing space through improved layouts or optimised processes, having less impact than a move or commissioning a new building. It is important to understand how the existing space is being used, how the additional space will be used and the predicted pace of the expansion. This final point is relevant because of the time it takes to commission new premises and therefore the need to predict the requirement for space at a point in the future when the new property will be occupied.

Demand

Figure 2.1, above, illustrates the different cycles of property supply against a typical corporate and public sector demand over time. Property has a very long wavelength, driven by the long-term nature of freehold ownership, average lease lengths of seven years and the length of time to deliver new premises, either because of build periods or the complexity of the transactions. Public sector organisations have a regular wavelength underpinned by budget and political cycles of three to five years. Corporate organisations have much less predictable cycles driven by market forces and the pace of organisational change. The graph shows the fleeting alignment between the amounts of space an organisation has and the amount it needs.

Consolidation

Consolidation may be the driver when less space is needed, either because of structural changes in the organisation or because opportunities arise that enable multiple sites to be consolidated. Examples include a need to bring staff or processes together from a number of locations: to a single site or building; following a merger; or because of internal reorganisation. There are two main advantages of this. Firstly is the potential to reduce the running costs of property, as it is more efficient to have one site than many, and if a building is owned, there is the opportunity to gain capital receipts. Secondly, there are operational advantages, such as easier communication and more efficient processes.

Location change

Location change can be either cost driven or operationally driven. The cost of buying, renting and running property varies across the country and the world, and therefore it is possible to change property overheads by moving locations. However, with the high cost of moving, the annual benefits of reduced rents may have a long payback period. Operationally driven needs – such as availability of suitably qualified staff, access to customers or distribution networks – can usually only be addressed by a move if the current location is judged to be wanting.

Opportunity

An opportunity may present itself that is too good to turn down – for example, an offer to buy the existing property from the owner or a chance to terminate a lease early. In this event it is important to see what else could be achieved by a move. A move from one property to an identical one in the same location may represent a host of missed opportunities to consider other locations, smaller, larger, cheaper or better quality premises, depending upon the organisation's priorities.

Quality improvement or cultural change

Quality improvement or cultural change occurs when properties age, maintenance costs become disproportionately high and eventually replacement provides better value for money. Keeping up with modern environmental standards and staff expectations cannot always be achieved within an existing building and commissioning a new building will enable those needs and expectations to be met. In a similar vein, if cultural change is high on the organisation's agenda, a move can create a physical manifestation of the new culture, and the property's power to support change programmes should not be overlooked or underestimated.

Question 3: What are the project constraints?

It is essential to consider organisational strategy as well as the constraints for the project, as these will have an impact on the options available. Typically, a project will be constrained by one of three interdependent factors: cost, time and quality. This means that only two of the factors can be delivered and you must choose which two.

Consider what the project needs to deliver: an

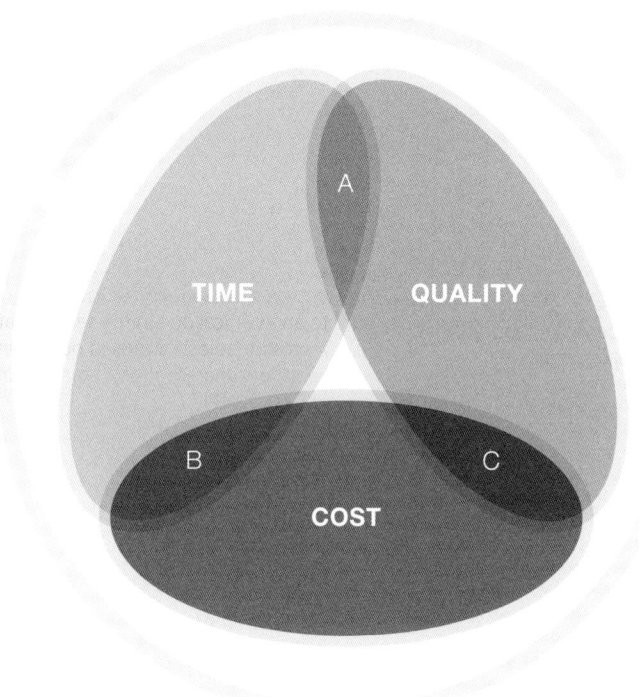

Figure 2.2: The interplay
between time, quality and cost

HQ building where a long-term investment in quality will reinforce the brand strength to staff and customers (Location A on the diagram); a sales office that needs to seize a new market opportunity (Location B) or an administrative office where costs need to be minimised and workers comfort maximised (Location C)?

As these three examples demonstrate, the constraints of a project can be expressed in terms of time, cost and quality, and their relative impact on the project must be considered to ensure that the correct balance is chosen to deliver value to the organisation. It is very difficult to deliver all three factors to the same extent. If quality is the top priority, then the project will need to be given time for the delivery as it is difficult to deliver quality at speed. Similarly, if time is not available to achieve quality, costs will have to increase in order to deliver the required quality on time. If there is a fixed budget for the project, then it is best if there are few time pressures to avoid compromises on quality that may fail to deliver value for money. These may seem obvious but unless the project team is made aware of the constraints and priorities, it may take decisions that are not aligned to the required project outcome.

Figure 2.3: The Vodafone headquarters in Newbury is designed as a number of separate but interlinked buildings on a single site. Vodafone currently occupies all the buildings but if its space requirement reduces for any reason, it would be able to vacate a whole building and let it to another occupier without needing to carry out any physical works. A self-contained building will also be more attractive to another occupier than a part of a large building with one dominant occupier.

Project influencers include the location of the property and the desirable levels of flexibility. Location will affect the price of land, construction costs and rental costs, but it may also influence the quality as if a project is over-specified for the location, the value-add may not justify the additional costs. Flexibility comes in many forms: design flexibility may enable extensions and subdivision for disposal as the Vodafone example shows.

Legal flexibility takes the form of options that enable contracts to be changed as the organisation changes, such as short lease commitment or ownership, allowing a sale or freedom to alter a property physically over time. Financial flexibility may be offered by leasing property, enabling a choice of affordability, or alternatively by owning a property, enabling money to be raised through borrowing or sale and leaseback.

Question 4: What options are available?

Now that you have assembled all the information about the organisation, the project vision and constraints, you can begin to consider which procurement route will deliver the best value for money from the range of options available. Options need to be considered within four categories:

▶ Tenure.   ▶ Design/location.
▶ Financial. ▶ delivery.

Tenure

There are broadly three ways to have an interest in land: freehold ownership, leasehold or under a licence agreement. Owning a freehold gives an organisation access to potential capital value increases but at the cost of initial cash investment. For this reason the return on the cash, if used elsewhere in the organisation, needs to be compared to the potential return and risks of a property investment in order to decide whether cash should be locked away in property. Unless the opportunity cost of the capital is considered as a measure of the cost of the property, it can appear 'cheap' compared to rental properties, which may skew comparisons of value for money or efficiency. Although ownership locks-in capital, it does provide stability, freedom from market fluctuations in rental costs and the ability to build or use the premises however the organisation wants, without recourse to any higher authority (other than planning legislation and building regulations).

Leasing properties has become more attractive in recent years, as freehold prices have risen out of reach of all but a few organisations and the lease agreements themselves have become more flexible. Leases typically range between five to 15 years and are readily available in most established commercial locations for standardised premises types such as offices, shops and warehouses. Because the property owner takes a calculated risk in readily finding a tenant when offering a property for lease, it is rare to find very specialist properties available on this basis, such as police stations, laboratories or schools. Leasehold properties require less and sometimes no up-front capital investment, depending upon the amount of physical adaptations required. Rents are rarely fixed for the term of the lease (unless it is for less than five years) and are usually increased every five years, in line with increases in the market rents for similar properties in the same location. This

can be difficult to budget and carries a risk to an organisation that rents will increase significantly, impact overheads and correspondingly have an effect on returns. It may be possible to negotiate rental uplifts that are linked to an index such as the Retail Prices Index, in which case rental overheads will be easier to predict and increases may be more in line with organisational growth. Leases offer less flexibility as to use, and physical alterations will require the landlords' consent. They also offer little scope to change size or the term of the arrangement unless options are negotiated when the lease is first entered into. If the organisation (tenant) cannot predict how much space it will need or for how long, then it should consider paying a premium for some form of flexibility, such as the ability to terminate the lease early. Taking a short lease would enable the organisation to move on if it needs to grow or shrink, but moving can be disruptive and there is no guarantee that the right space will be available when it is needed.

Licences for occupation are typically used where the term of the agreement is short, ie less than a year, and usually for smaller spaces. They would most commonly be used for serviced offices or workshops and will often include the provision of services by the landlord and very little responsibility on the part of the tenant. They offer an easy and fast alternative to traditional leases and could be useful where the goal of the project is to provide space on a short timescale.

Financial options

Organisations have the choice of owning or leasing property according to their financial profile. Owned property will have a significant upfront purchase cost but with no rent to pay, the impact on revenue budgets is low. However, in the case of a purchase, the opportunity cost of the capital needs to be considered for the true cost to be understood. This is particularly true for an organisation following a strategy of expansion where capital will drive the greatest returns from investment in the business rather than being tied-up in property. However, leasehold premises may still require some upfront investment for alterations that will be a visible and possibly substantial operating cost for the organisation. When/if the accounting standards change, leases will have an impact on the balance sheet as they will have to be capitalised and an organisation sensitive to balance sheet changes will need to consider this.

Design options

What is this requirement actually for? Is it a commoditised product, such as offices, or specialist – such as laboratories? Reference to the Town & Country Planning (Use Classes) Order 1987 is useful, as this categorisation of property uses is referred to throughout the property industry. Premises such as shops, offices and warehouses will be readily available nationwide and in most developed countries. Therefore, the number of choices in terms of tenure, location, cost and quality should be large. However, properties suitable for use as, for example, care homes, hotels, schools and medical centres will be more restricted and it is likely that some physical alterations will be required to an existing property, with fewer tenure and location choices available. For very specialist uses, such as hospitals, laboratories, prisons, blue-light and military facilities, location, there may only be one or two tenure and delivery options, which will impact on time and cost.

As well as choosing a geographical location, it is important to consider the position in the local area, such as town centre, shopping centre, out of town business/retail park or countryside. Different choices will have different price profiles, staff attraction and retention factors, availability, tenure and quality options. For example, a standard office requirement placed in a town centre will provide good transport and welfare facilities for staff without any additional investment. The same office requirement placed in a business park may require additional parking to make up for the lack of public transport and possibly on-site catering facilities – which will add to costs. A shop on the high street may be available freehold or leasehold but it might be in an old building that could have cost implications for the fit-out and maintenance. A shop in a shopping centre is only likely to be available on a leasehold basis and landlord charges for shared customer facilities such as parking, security and cleaning will be an additional cost. However, the retail unit is more likely to be modern, in good repair, regular in shape, easy for service access and attract more custom.

Delivery route options

There are basically two delivery options: to take an existing property or to build a new one. Under each of these there are a number of other options. A suitable existing property may enable the organisation to move in without any physical works being necessary, such as a serviced office where carpets, power, data and even furniture are supplied ready to use. However, it is more likely that some degree of works will be required, either to alter the layout or to improve the quality of the finish or services. This will require some capital expenditure unless, in the case of a leasehold agreement, the landlord is able to undertake the works and reclaim the cost through a higher rental.

When it is decided that a new building is required, there are in turn further options available. Sometimes land is offered by a development company who is able to manage the building of a facility either for a lump sum capital payment or in return for a commitment to take a lease of the completed property; the obvious difference being the capital versus revenue commitment. The advantage of commissioning a building from a developer is its experience in the specifying, pricing and delivery, which will reduce the risk of time and cost overruns, and provide quality assurance. However, depending upon the level of risk the developer is taking on cost and time – ie to what extent it has guaranteed either to you – this may represent good value for money but will not necessarily be the cheapest option. For public sector organisations the option of the private finance initiative (PFI) is available, where

a private company (usually a contractor, developer or facilities service provider) will design, build and then maintain a facility in return for a long lease from the organisation. This route provides the organisation with cost, time and quality certainty and all-important capital funding. However, it has been criticised as a procurement method as it restricts the organisation's ability to change the premises physically or to alter the terms of the contract if it's circumstances change, resulting in some public sector bodies being saddled with premises that no longer offer value for money, as they do not deliver what is now required by the organisation although the cost remains fixed. The latest version of public private partnerships is PF2, where the government retains a share of the equity and therefore more control over the project, as well as a share in the returns.

The final option to be discussed here is really the purpose of this book – to guide the organisation through the process of specifying, designing and constructing a building for itself. Although this may appear at first glance to be a cheaper option, as no profit is paid to a developer or landlord, in return the organisation must take on the risk of the project taking more time, costing more or not being fit for purpose, all of which may have significant impacts on the organisation itself, with budget allocations, disruption, further works and ongoing maintenance issues. Key to making this option work will be selecting the right team of experts to advise on the various elements and stages of the project. This book will help you with that process.

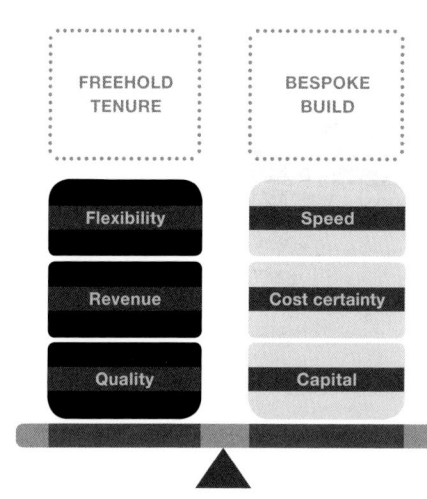

Figure 2.4: The organisation buys or already owns the freehold and commissions the construction of a bespoke building

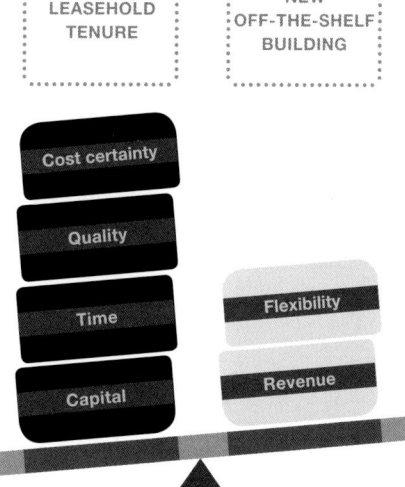

Figure 2.5: The organisation takes a lease on a new or refurbished building, owned by an investor or developer

Question 5: Which option should the organisation choose for this project?

It's time to review what you have established so far:

▶ What is your organisation's strategy? Is it growth, improved product quality, improved margins, reduced overheads or culture change?

▶ What are the drivers for this specific project? Do they include expansion, consolidation, location change, opportunity or quality change?

▶ What are your project constraints? Do they include time, cost or quality?

Note what is important to your organisational strategy, project driver and constraints from the following:

▶ Cost certainty.

▶ Fast or certain delivery time.

▶ Low capital requirements.

▶ Low revenue requirements.

▶ Design or contractual flexibility.

▶ Finish or location.

▶ Quality.

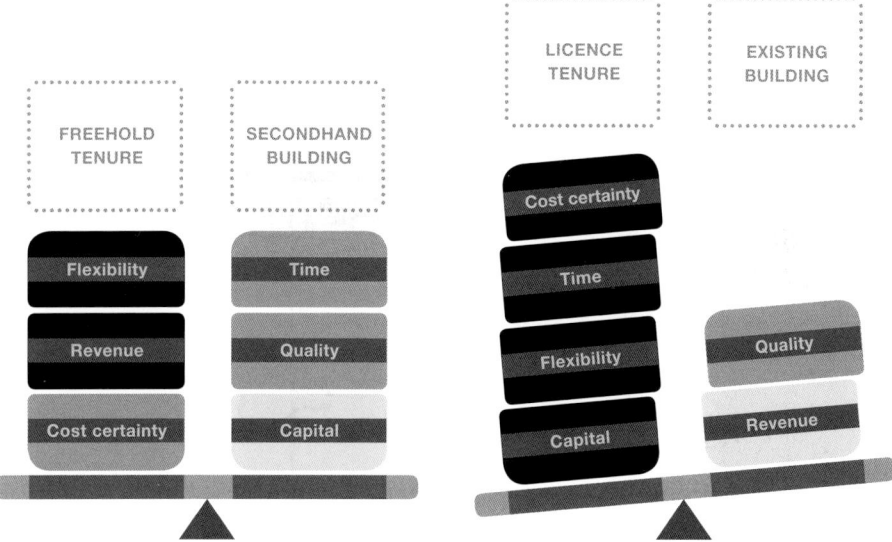

Figure 2.6: The organisation finds an existing building and buys the freehold. It will then commission any necessary alterations

Figure 2.7: The organisation takes a licence of a suite of offices or workshops from an owner or operator of a multi-occupied building

Check your selections against the following groupings, this will help you to decide which property solution might best fit the organisation. Black is good and white is bad.

If you are comfortable with the risks and rewards of the bespoke building solution, then you are ready to move forward. If, however, another procurement route, such as an off-the -shelf acquisition would suit you better, then this should be explored before you commit the organisation to commissioning a new, bespoke building. ■

What makes a good client

Commissioning a building project and having it meet your aims requires consideration of multiple factors, many of which often operate in apparent conflict with one another. The factors for the client to consider and the motives driving them might be represented as illustrated in the table below (and even then that is materially to simplify matters).

The message to be taken from table 2.1 below is that there is a balance to be struck between time, cost and quality while also balancing the client's own requirements with those of enhancing the built environment – and having a responsibility to the public at large in terms of such factors as sustainability. In the case of those clients who receive public funding for their projects, there will be the further overlay of public procurement directives to be considered.

Motive	Pros	Cons
Building to be as cheap as possible	Minimises capital outlay	May produce an eyesore (inconsistent with making a contribution to the built environment, risking corrosion of the user's brand) and also incur higher running costs (which not only hits the client's revenue stream on a recurring basis but is inconsistent with responsibility responsibility regarding sustainability)
Building to be completed as early as possible	Building may be put to use at the the earliest opportunity, which is particularly relevant for commercial projects where income generation is involved	Quality may suffer. There may also be increased construction costs involved in compressing the programme
Building to be completed to the highest quality of design, materials and workmanship	The building may make a contribution to the built environment, be sustainable and provide a conducive facility in people to work in	It may cost too much and take an uneconomically long programme to realise! Moreover, with technology at the forefront, the project may be particularly susceptible to variations
Design led brief/trophy architecture	The client's requirements as to quality will be met	There may be a cost premium and lengthy programme as a result
Budget led brief	The capital outlay will be in accordance with expectations	The quality and functionality of the completed building may compromise the intended use of the building

☐ Table 2.1: The implications of various procurement strategies.

Perspective
by Mark Hackett

An effective client will recognise that in balancing what are often competing objectives there is an optimum arrangement to be struck. This requires, first of all, great thought to establish and articulate and then, secondly, regular management to maintain. The process can be made easier by having value management workshops in which the actual – rather than perceived – benefits of particular building features can be tested. If necessary they may be excluded from the brief. There is a distinction to be made between that which is indispensable and that which is simply desirable (though by all means have that which is desirable if your budget allows).

Given the tension between budget and brief, it was once said of Aston Webb, the Edwardian architect, that he would say to his clients: 'Tell me what you want or what you can afford but please don't tell me both.' Hopefully value management techniques allow one to bridge the gap, thus following the invocation of a great figure from another sphere, namely the Nobel Prize winning physicist Sir Ernest Rutherford (1871–1937) who said: 'We haven't the money so we've got to think.'

Even in the best planned of projects, disputes may arise. As soon as you insist on a specific commitment (eg 'this is the building I want' or particularly if you go further and say 'and I want it completed by a specified date and for a specified price'), the seeds of confrontation are inevitably sown. It is not forms of contract or methods of procurement that are of themselves confrontational; it is the parties to a contract who, having been required to undertake commitments, create the confrontation by either failing to achieve them or actively seeing to relieve themselves of them. Certainly, ill-prepared contract documents, or wanton changes in them thereafter, will produce much fertile ground in which claims will propagate. But to seek to blame particular forms of contract or methods of procurement for the widespread recourse to claims is to tilt at windmills, with the result that the real enemy is ignored.

Even in the best-planned projects, variations or changes in design will be required and it is unrealistic to attempt to ban them. After all, the process of construction usually involves a unique, complicated project erected largely in the open, on unpredictable ground, and in more unpredictable weather conditions. There can be a considerable time span between making the initial decision to build and the completion of works on-site. During this period, there will inevitably be changes in technology, fashion, regulations and the client's requirements.

The initial focus should be on developing the brief so that the design team knows exactly what is required and is not left to guess a requirement whose correction will come late and at great cost. Getting the brief right from the outset using careful and precise orchestration will pay long-term dividends. The members of a design team possess myriad skills and the client, like a conductor, will have to draw out what is needed at the right time. To be able to do this requires an understanding of construction procurement and techniques, and even lay clients should invest some time into reading around the subject if they are able to integrate with and get the best from the design team.

Chapter 3 Defining the project

by Joanna Eley

"It is very important to get the early version of the brief right, as the more detailed versions will be built on it. Setting off in the wrong direction is a major reason why a project may be unsuccessful."

Introduction

There is very rarely a single solution to a problem or requirement that leads to a new construction project. This chapter looks at defining the chosen solution though its brief. It covers:

▶ Developing the brief in increasing levels of detail.

▶ Who is responsible for the brief at different stages.

▶ Factors that affect how a project can meet the objectives.

▶ Stakeholders and how to manage them.

▶ How much risk for the sake of innovation – is it worth it?

▶ Learning from the project; the role of post-occupancy evaluation.

Project definition

There is a project if the client's outcome can be achieved by a building project. The first role of the brief is to identify the desired outcomes and then describe, in increasing detail and at each stage, the project that will meet them and how it will be judged to have done so. So, at the start of a project the client has a major responsibility and the best opportunity to clarify the desired outcomes and the criteria for success. When a brief is unclear, incomplete, inappropriately detailed for the stage of the project, unrealistic in terms of time, cost, or other controlling factors, the project is at great risk of being unsuccessful. This chapter is therefore critical for the client to

appreciate where they should have an input and when to get support for matters beyond their abilities. A business case is an equally important component of the project and the planned outcomes provide the foundations for this.

The 'brief' defines what and for whom the project is for. The brief is the key communication document for the entire project and gradually evolves from an aspirational statement to a full and detailed description of what is to be built. A number of different names are given to various stages of the brief. The terms below are used here, but regardless of terminology, clarity is essential in setting out the type of information required and who provides how much detail at different stages.

Stages in developing a brief

1 The vision statement

2 The strategic brief

3 The project brief – which is made up of:

4 The initial project brief

5 The design, technical or final project brief

Typical patterns of activities associated with these stages of the project are shown in the table, below, alongside the first stages of the RIBA project cycle.

Supporting activities		RIBA Plan of Work 2013 stages
• Project visits • Special studies • User needs • Business case	• Vision • Strategic brief • Initial project brief	Stage 0: Strategic Definition
• Stakeholder consultation • Special studies • Communication plan	• Design brief • Sketch designs • Final/technical brief	Stage 1: Preparation and Brief

Table 3.1: Activities related to these RIBA Plan of Work 2013 stages

New build or refurbishment

Although it may seem that a new building is more complex than a refurbishment project, in fact many – often very large – projects involve work to existing buildings with their own features. Established situations may be as complicated as new sites. An existing building is in effect the 'site' for such a project and introduces unknowns (eg is there asbestos to be removed?), complexities (eg how to work around existing occupiers and the organisation's 'work as usual') and problems (eg decant

space being taken up by expansion faster than planned). Appropriate time and skills must be deployed to understand the the impact of achieving the desired outcomes.

Spend enough time at the start

Valuable time spent at the start of project is when the direction is set and the outputs targeted, perhaps investigating other buildings, talking to other clients, starting to develop a language to be shared with others on the project and clarifying options for different ways to approach the perceived problems. It is very important to get the early version of the brief right, as the more detailed versions will be built on it. Setting off in the wrong direction is a major reason why a project may be unsuccessful.

Developing a brief

If the decision has been made to build, the next stage is to understand the implications of the project, such as cost in financial outlay and in client time, time frame, disruption, risk etc. Some design input to establish the likely size and type of project and financial input from a quantity surveyor based on broad design ideas will be needed to assess feasibility.

The earliest brief is the seed from which a project can grow. This happens before a building project has even been decided upon (Stage 0 of the RIBA Plan of Work 2013) and could simply express the client's vision for the project outcome. The client's vision needs to be able to endure, direct and inspire all these involved. The desired outcomes must be clear, agreed by relevant decision-makers and simply expressed early on to help prevent 'mission creep' during briefing and then during the project.

A clear decision on whether to embark on a building project must be based on a good understanding of which aspects of the building will help deliver the necessary outcomes, and what else – for example in management approach or use patterns – will also be important. Some of these other aspects may need to be progressed at the same time as the building work.

Who creates the brief?

The client's role in creating and developing the brief starts with taking full responsibility for content. This responsibility diminishes as consultants and specialists take increasing authority for producing it. Nonetheless, the client must understand and agree the final version of the project brief, on which the design will be based and the project built.

Vision or statement of need

At the start of the journey (RIBA Stage 0) it is often an individual, probably representing a wider client group, who starts the briefing process by making a 'statement of need', or perhaps making 'a vision statement'. A statement of need could say: 'We are fast running out of office space and need to increase our space by 20% in the next five years.' A simple vision based on this might, for example, say: 'More office space will allow us to provide the most competitive and effective service to our customers and a 21st Century working environment for our staff, allowing them to achieve a better life-work balance.' Embedded

in this simple statement are a number of desired outcomes beyond 'more office space' to be more competitive, serve customers well and provide staff with excellent modern space and a better life-work balance. The strategic brief provides a fuller understanding of the hoped-for outcomes and the way in which a building can be expected to help support this. Realistically, a strategic brief is often in place before the final 'vision' that would look beyond this to aspirations for which solutions still need to be imagined, formulated and distilled, and shared with all involved.

Strategic brief

At the next stage (RIBA Stage 1), the client is likely to need help to fully develop their strategic brief that sets out the desired outcomes of what the project must achieve, reasons why this is needed and any significant known constraints, especially in respect of location, budget and timeframe. The first pass at a strategic brief may sometimes show that specialist help will be needed before a fuller project brief can be prepared, so the strategic brief should also be aimed at providing the information needed to appoint advisers and consultants.

When assembling a clear strategic brief, an RIBA Accredited Client Adviser experienced in brief writing can be of great help. There may even be one involved already, who has helped to establish how the outcomes sought will be best achieved through a building project. A cost consultant is also likely to be needed. Others, such as planning, landscape or heritage advisers, will be important at the next stage when the initial project brief is developed, and may sometimes be needed to help with the strategic brief for some projects with relevant special features. Members of the client's in-house team with strategic understanding of future needs may also need to participate in developing the strategic brief to ensure that the project gets off to the right start, avoiding time wasted on 'blind alleys'. The client can continue to develop the brief using feedback from advisers until it describes the requirements in enough detail for feasibility studies and/or option appraisals to be commissioned and prepared. Once the required project has been defined in more detail, the initial project brief can be written.

Initial project brief

The initial project brief should make clear the extent to which it is desired, financially possible or relevant to the outcomes to seek something iconic and unique, or whether something more standard, functionally effective but unremarkable is needed. The client may prepare it with or without specialist input. An RIBA Accredited Client Adviser is able to give guidance, especially in terms of the language used where a misunderstanding with the design team can lead to later problems.

This brief now becomes the key document on which all design decisions are based comprising:

▶ All existing information, such as the business case, strategic brief, etc.

▶ Site information – surveys, appraisals, etc.

▶ Additional stakeholder information from workshops, interviews, surveys.

▶ Statutory authorities' input – fire brigade, statutory utilities, local authority, heritage.

It will generally include:

Stage 1 A description of the client – individual, or corporate

▶ Vision, mission, organisation and culture.

▶ Changes that the project will bring about.

▶ Interface with any other on-going or planned projects.

▶ Client policies and preferences that will have an impact.

▶ The principles to be followed in developing the design – such as the importance of sustainability, or minimum disruption to.

Stage 2 Project and construction requirements/constraints

▶ Planning constraints.

▶ Site topology, ground conditions.

▶ Budget – with respect to running as well as capital costs.

▶ Programme, key milestones.

▶ Known risks to be developed into a risk register.

▶ Consultation outcomes.

Stage 3 Spatial requirements – leading to an accommodation schedule/space budget

▶ Number and type of users, adjacency requirements, likely changes in future.

▶ Areas required including for specialist needs.

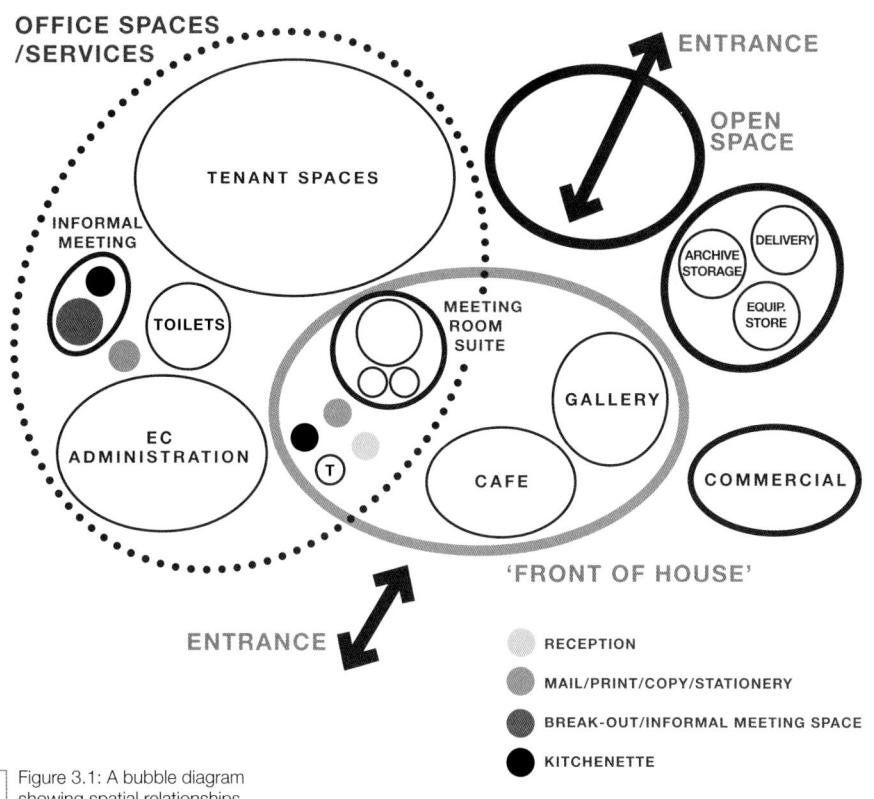

OFFICE SPACES /SERVICES

ENTRANCE

OPEN SPACE

TENANT SPACES

INFORMAL MEETING

TOILETS

MEETING ROOM SUITE

ARCHIVE STORAGE

DELIVERY

EQUIP. STORE

GALLERY

EC ADMINISTRATION

T

CAFE

COMMERCIAL

'FRONT OF HOUSE'

ENTRANCE

RECEPTION

MAIL/PRINT/COPY/STATIONERY

BREAK-OUT/INFORMAL MEETING SPACE

KITCHENETTE

Figure 3.1: A bubble diagram showing spatial relationships

- ▷ Zoning, security/separation requirements.
- ▷ Circulation guidelines.
- ▷ Relationship of internal/external spaces.
- ▷ Space data sheets.

Stage 4 Technical requirements

- ▷ Structural strategy.
- ▷ Servicing assumptions (MEP and others) and sustainability targets.
- ▷ Fire compartmentation assumptions.

- ▷ Facilities Management (FM) issues, eg maintenance, security, waste, etc.
- ▷ Component issues, eg long lead times, need for specialist design, etc.

Stage 5 Special issues – for example

- ▷ Heritage considerations.
- ▷ Site complexities.
- ▷ Stakeholders likely to be concerned.
- ▷ Other – specific to the circumstances.

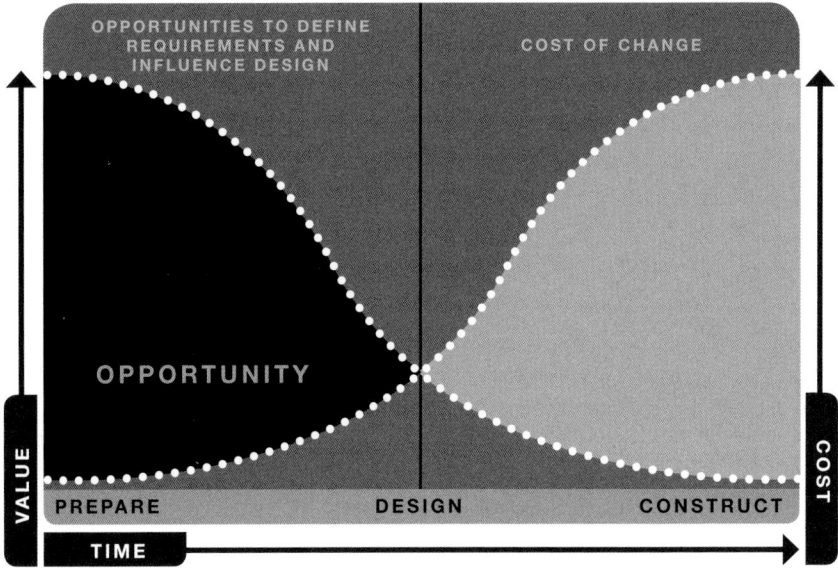

OPPORTUNITIES TO DEFINE
REQUIREMENTS AND
INFLUENCE DESIGN

COST OF CHANGE

OPPORTUNITY

VALUE

COST

PREPARE   DESIGN   CONSTRUCT

TIME

Figure 3.2: Time spent at the beginning provides
opportunities, late changes cost money

Design brief

The initial project brief evolves and becomes
more detailed during the early iterative stages
of the design process (RIBA Stage 2) turning
into a design brief. The fundamental direction,
however, should not change unless a different
strategic approach has had to be adopted and
is agreed.

The consultant team will develop the brief as
design ideas are elaborated, discarded or new
ones emerge to meet the project objectives. It
must be fully understood by the client and then
agreed and signed off by them. At the end of
RIBA Stage 2, the design brief should be fixed.
Further development of the design will take
place before construction, based on a fixed and
agreed design brief to minimise risks and costs
associated with changes after this point.

As the project progresses through detailed and
technical design (RIBA Stages 3 and 4), the brief
may become extremely detailed, for example
incorporating room data sheets for every space

prepared by professionals and the client who
should agree and sign them off before the
next stage of the technical design starts. When
complete, the client signs off the design brief.

Achieving success

The interaction of many factors produces a
successful project, meeting its objectives and
giving the client, the users and the public
something even better than they expected.
Some of the main issues and questions to be
considered are listed below. Answers affecting
each other need balancing to understand risks.

Stage 1 Has a proper brief been prepared?
Have criteria for success been established? Have
all the right people been involved and agreed
with these? Without this there is little chance of
achieving the desired outcomes.

Stage 2 Is the budget realistic? Is there
enough money and can it be released at the
right times to progress the project though all its
stages? Has the internal cost of people's time
been allowed for?

Stage 3 Has enough time been allowed – especially at the start? This is not just for the construction period. Was enough time given initially to deciding that the right broad direction was chosen – the best option, the right level of innovation? Has a realistic amount of time been allowed for design? Have all members of the client team been allowed time to participate properly? A client adviser can help understand what level of time commitment this may mean.

Stage 4 Is the team right? Does it have an open and positive approach to collaboration with each other and the client? Does it have the skills the project requires? Has it been selected on this basis rather than on cost? Are all the individuals committed to delivering a building that the client will be able to manage, adapt if needed and enjoy? Adopting a Soft Landings approach is recommended to achieve this.

Stage 5 Is the procurement route appropriate? Is procurement dictated by a procurement department or has the route decided on been matched closely to the risks and desired outcome of the project? If the client team is not in a position to gauge this, it is another area in which a RIBA Client Adviser can help.

Stage 6 Is the chosen site (or existing building) suitable? Are known risks accounted for in the time and cost? Are suitable actions being taken early enough to reduce or remove risk? These may include early consultations with planning and other authorities, surveys and examination of existing conditions, and stakeholder communications. Advice can be sought on which early actions are needed and how early in the process.

Stage 7 Are all the stakeholders happy? This is explored in more detail below.

Managing stakeholders

Projects can be important to non-client groups called stakeholders, including funders, users, local people and a wider public interested in a site, particularly a public site like a park, the renovation of historic buildings, or a site of historic, environmental or other special significance. Some stakeholders, such as users and local people, may expect to be consulted, whereas relevant statutory authorities have a right to be consulted. A stakeholder management plan will define the consultation process, particularly important when consultation is critical to the project.

Balancing the risk of innovation?

When producing iconic buildings or developing innovative solutions, the implication is that there are few, if any, precedents. Unforeseen issues may pose a risk to the timetable, the budget or the overall outcome. While all buildings are in some sense unique because they are site-specific, there are many situations when tried and tested solutions can be put forward with only minor modifications to suit a particular situation. When this can be done, there is considerably more cost and programme certainty, and a high chance that the design – and perhaps also the construction – team have made similar buildings and thus have resolved some problems in advance. What this approach cannot do is meet unusual or specialised needs, deliver more than the client body or other stakeholders had hoped for, provide a distinctive brand or image, work with unusual sites and complex planning contexts, or make a new and different 'statement' building to meet a bold and forward-looking vision. A building to fulfil this solution is not always needed, but neither is a bland, lowest common denominator solution.

It may be important to convince board members, governors or other decision-makers of either side of the argument. They may need to be persuaded not to seek an icon when that will do little to help achieve the desired outcomes and also when it can be seen that inherent risks are likely to jeopardise important requirements. Equally, it is important to see potential benefits to a solution that goes a bit further than the leanest, simplest output.

The right choice for the project depends on understanding the ways that a little more input could produce a greatly improved outcome, and what types of risk cannot be accepted. For this decision, which is needed early on as it becomes embedded in the vision, all levels of the brief support from a client adviser may be helpful. A good way to get to grips with the issue is to visit and discuss a number of built examples, both adventurous and unadventurous, and to understand the merits of both. It is important to try to get facts about costs and other aspects of the process to identify how far any particular direction represents an unacceptable risk to the project in hand. Most well designed and constructed sit comfortably between the innovative and the banal.

Learning from experience: the role of post-occupancy evaluation

For clients that undertake a number of building projects in sequence, Post-Occupancy Evaluation (POE) can be carried out in many different ways. POE looks to answer a number of questions. It is a way to check that a building is performing in energy use and other service elements as intended, as well as looking at the design and build process to understand what could have been done better. Did the type of procurement meet the needs of the project? Did the way the team was chosen provide a good group that was able to carry out the work to the desired level? Could communications and trust between team members have been improved and would this have avoided problems? Did the brief capture the right information about user and client needs?

The answers to such questions will help get things done even better next time. There are still POE benefits for one-off projects. One such benefit is to understand anything that could be improved for the current building once it is finished.

The key to POEs done before or after a project is to be clear about the questions that need to be asked, collect real evidence through observation, interviews, measurements to answer them and interpret the information in a way that will help make decisions to improve the project and any that may follow it. ■

Clientship

About 90% of those who commission construction do this only once or rarely in their lifetime. In stark contrast, experienced clients are responsible for a significant monetary value of the construction industry's work, where 20% of clients commission 80% of the workload.

There are two types of experienced clients:

1 Those who build often but produce specialist facilities each time, eg universities which may build a library and then a hall of residence.

2 Experienced clients such as supermarket chains that produce the same type of buildings.

Experienced and inexperienced clients tackle briefing in different ways. Clients who build regularly have well developed briefs for those elements of their facility that are often repeated. Clients who build infrequently have to generate all elements of their briefs from the start. For these, it is important to learn about the elements of good clientship.

My definition of designing is preparing a plan for change. In consequence, everyone is a designer, whether planning an enjoyable evening for friends, designing a new insurance product, or robbing a bank. We need to help clients to understand that they are essential members of the design and construction team in order to help define the brief at different stages of the project. The brief is not a one-off document setting out the client's needs at the start of the project, but a continuous and developing process from inception to completion. For a successful building, briefing has to be an increasing interaction with designing, and subsequently also with construction. For that, the client's continuing integrated membership of the design and construction team is essential.

Once the client has accepted this logic as a precondition for achieving a successful building, the design and production of the building becomes a most enjoyable collaborative process, enabling each member of the team to learn from each other.

This process brings the widespread abandonment of what are assumed to be 'specialist' fields of knowledge and activities. Health issues become the exclusive domain of health professionals, legal issues of lawyers, technical issues of specialist technicians, financial issues of accountants and finance institutions, and so on. Instead, a different frame of mind has to develop, which shares knowledge to best advantage and recognises the true function of architecture and the development of our urban environment as an important cultural and social activity for clients, building users, building designers and building constructors alike.

The management of risk, for example, becomes a highly creative part of the process, instead of being relegated to exclusively legal and financial considerations.

The overriding skill required of the client is that of an intelligent human being sharing the challenges and joys of designing and building with the rest of the design and construction team.

In turn, this brings about a very different view of the built environment, enabling the client to reclaim a thorough understanding of the cultural, social, physical and economic determinants of our built environment.

Chapter 4 'How could my project be procured?'

by Ben Hughes

"A client organisations' opportunity to select an effective procurement strategy that best fits with its own capabilities and has the highest probability of delivering successful outcomes is greatest at the outset of a project."

Introduction

This chapter describes procurement steps and options for an individual project or programme that delivers a new or enhanced asset (referred to as either 'project' or 'projects') for a client organisation. The 'Improving Infrastructure Delivery: Project Initiation Routemap Handbook'[1] from HM Treasury defines procurement as:

'How to engage with the market, determine optimum allocation of risk between the client organisation and the supply chain, package up the work to be procured and identify the most appropriate procurement route and form of contract.'

In the first instance, understanding the designation of a client organisation that is funding a project (between either public or private sector) can dictate the array of controls associated with a procurement process, resources and structure of the client organisation. For example, with a privately funded project such controls would include internal and external audit requirements and a review by funding partners, whereas a publicly funded project will also need to consider compliance with procurement regulations[2] (in order to comply with European Union procurement directives[3]), in addition to adhering to more specific central or local government requirements.

It is a common belief that mandatory adherence to the Public Contracts Regulations[4] (referred to as the 'regulations'), underlying European Law and Council Directives generally results in a longer procurement process. However, it is also worth analysing the disruption caused by incomplete or poorly defined requirements supplied by the client organisation together with changing business-critical outcomes rather than purely blaming the process.

Irrespective of the actual designation of a client organisation, when procurement is treated as an integrated function within the other project team disciplines it can be an effective tool, provided it considers the following underlying structure:

1 Analysis: an explanation of the nature of the challenge, condensing complexity into a simpler understanding of key issues that must have clear alignment to attaining each project's business-critical outcomes.

2 Strategy: an overall strategic procurement approach chosen to cope with or overcome the obstacles identified through the analysis that can flex in response to any internal or external influences.

3 Actions: coordinated procurement activities that are integrated with the overarching project programme that supports the accomplishment of the strategy.

Of course the relative complexity and size of each project, together with the nature of business-critical outcomes, will dictate the depth and level of intricacy required to develop and implement a successful procurement strategy.

Why should procurement be considered at the start of the project?

A client organisations' opportunity to select an effective procurement strategy that best fits with its own capabilities and has the highest probability of delivering successful outcomes is greatest at the outset of a project ie the appraisal stage (Stage 0 of the RIBA Plan of Work 2013).

As a project develops, apart from the passage of time, inevitably decisions are also required that can influence – sometimes unintentionally – the achievement of business-critical outcomes. The later a procurement strategy is properly considered in the work stage cycle, the less effective that strategy becomes as a tool and more likely it would have the characteristics of a mechanistic, process-led activity that would add little value to achieving desired business-critical outcomes.

Procurement considerations

The following features all require consideration in the lead up to a procurement process and continuous assessment during the process. Client organisations will need to prioritise and balance these according to their particular business-critical outcomes, organisational structure, capability, competence, regulation and budgetary restrictions.

Achieving project quality

▶ Should not be assumed nor taken for granted as a consequence of selecting a particular procurement route, and the client must emphasise and define the quality levels required throughout the process.

▶ Takes account of appropriate compromises needed to be made between time and cost in the full knowledge of any consequential impact on quality to be delivered.

▶ Requires provision to be made for innovation (virtually every building is a prototype) and how to allow differentiation between offers from potential contractors.

▶ Should take account of wider factors including whole-life considerations, choice of materials and products and recognising that the long-term operational cost for an asset is likely to far exceed the once-off capital cost of creating the asset.

Achieving cost certainty

▶ Depends on the probability of completing a project's scope within the budget agreed between a client and contractor before the commencement of construction.

▶ Relies on a client organisation's own cost benchmarking capability analysis and experience that might influence or undermine cost certainty.

▶ Is influenced by budgetary and/or regulatory or legal restrictions.

Achieving time targets

▶ Relies on the reliability of completing projects on time compared with that planned and

▶ Requires the identification of internal and

external factors that could cause delay or disruption to the schedule, eg third party consents, product or service availability (scarcity), approvals and governance.

Dealing with technical complexity

▶ Depends on the client organisation's prior experience in delivering a project of the type under consideration and the impact this has on:
- Risk allocation between it and the contractor (s).
- The shape of its delivery organisation.
- The clarity, precision and manner of specifying the client's requirements.

Providing the flexibility to accommodate change

▶ Requires the client to have an understanding of the level of flexibility required to allow for changes in requirements at various work stages.

Understanding and managing project risk

▶ Requires clients to grasp the extent of uncertainty of funding, their organisation's requirements and the nature of any constraints or dependencies in these areas.

▶ Needs clients to assess the amount of foreseeable changes and the definition of processes or allowances that should be aimed at securing best value during the procurement process.

▶ Needs an answer to the question of who is best able to own and/or manage the risks:
- Client.
- Contractor.
- Others?

▶ Needs clients to ask themselves whether they are able or willing to carry risks and

▶ Demands procedures for ownership, monitoring and mitigating risks aligned with any commercial principles.

Complying with statutory constraints and considerations

▶ Needs clients to be aware of the application of the regulations and, if so, what procurement procedure is best suited to the project? (See below).

▶ Requires clients to be aware of any specific planning requirements and constraints ie Section 106 Agreements, development consents, listed building consents, or similar conditions and restrictions on development.

Procurement timing depends on client capability

▶ Identifies the pre-contract organisation including all procurement, legal and subject matter experts, senior management stakeholders, approval and corporate governance processes.

▶ Achieves an early identification of the post-contract organisation and devising appropriate interventions to improve, enhance or augment capability.

Liaise and manage the handover between the pre-contract and post-contract teams during the transition to the post-tender phase of the project.

What is best practice in procurement?

When procuring projects, there are certain principles that combine to underpin best practice and these are broadly transferable across all projects. Given the range of considerations associated with the procurement of projects, best practice has many facets and broadly seeks to satisfy the following:

Takes account of the design and execution of a procurement strategy based on the client's capability and the project's complexity.

Enables the contractor and its supply chain to take a margin reflecting prevailing market conditions.

Delivers the requirements of external stakeholders whether these are regulatory, community-related or others, eg funders.

Ultimately delivers specified business-critical outcomes.

What does best practice look like?

The management systems that underpin examples of procurement best practice include:

Structured procurement processes

Providing a clearly defined, unbiased process for the selection of suppliers (including transparent evaluation systems and criteria) that determines the ability of suppliers to

meet the specified requirements[5] and add value.

Quality-based selection processes

These can enable the identification of potential suppliers with relevant experience, and selection of the most appropriate firm on a qualitative basis, thoroughly discussing the project details with the best-qualified firm.

When applied to larger projects, suppliers should be pre-qualified based on (in part) their implementation of best practice, where responsibilities are written into tender and contract documents.

Offer non-adversarial selection methods that do not force costs down compromising resources such as staff and materials.

Means that the most important quality-based attributes by which a supplier's suitability to carry out a particular project can be judged regardless of the selection process are:

- Technical competence (including safety).
- Managerial ability.
- Availability of resources.
- Corporate integrity.

Achieving best value

Is achieved by choosing the right type and level of competition.

Is gained by adopting processes where the design and construction teams can and should work together as an integrated team.

Means taking a 'whole-life' approach to specifying and costing the project.

By regarding value chain evaluation as a multi-value system where each of the stakeholders evaluates value from its particular perspective.

Why use best practice procurement procedures?

The main benefit derived from best practice procurement procedures is improved returns from increased asset value and reduced whole-of-life costs when measured in both monetary and non-monetary terms. This may be achieved through:

- Reducing costs associated with maintenance, building re-fitting and safety risks over the economic life of the asset.
- Improving design quality and innovation as pressure to minimise costs is replaced with positive incentives to achieve best practice.
- Increasing market sustainability as the pressure on suppliers/providers to simply cut costs is reduced and they are encouraged to innovate and seek better design solutions.
- Improving risk management by allocating risk ownership appropriately to those most able to influence or control them.
- Making a positive impact on the community, the environment and other external stakeholders.
- Reducing the likelihood and number of contract disputes.

Poorly defined requirements or mechanisms (not necessary understood by all the key parties to the contract) and errors or omissions from the contract documents all feature in the top five causes of EC Harris' latest global construction disputes report in Table 4.1 below:

2011 Rank	Cause	2012 Rank
New	Incomplete and/or unsubstantiated claims	1
New	Failure to understand and/or comply with its contractual obligations by the employer/contractor/subcontractor	2
1	Failure to properly administer the contract	3
3	Failure to make an interim award on extension of time and compensation	4
2	Errors and/or omissions in the contract document	5

☐ Table 4.1: Global causes of dispute

All of the above revolve around a mistake or failure which makes them all to some extent avoidable. In the UK, more recently, EC Harris reports that it has become more prevalent for disputes to occur due to all parties failing to understand their contractual obligations, which has been linked to the ambiguous and sometimes overly legalistic drafting of bespoke contracts. Therefore, if client organisations adopted more standard forms with fewer amendments, the occurrence of this type of problem could be reduced.

Recent developments – collaboration in construction

▶ HM Treasury – UK Charter 2011, establishes the high-level objectives and behavioural changes needed to reduce the costs of infrastructure delivery. These objectives are set out in the commitments made in the Government's Plan for Growth at Budget 2011, the Infrastructure Cost Review: Implementation Plan (March 2011) and are consistent with the Government Construction Strategy (2011).

▶ BS11000 – Collaborative Business Relationships

- Businesses in the construction industry have come to recognise that promoting collaboration and achieving less adversarial working relationships between clients and contractors helps attain business-critical outcomes.

- In 2010, BS 11000 was published in two parts:
  ▷ **Part 1:** A framework specification, defining the standard against which organisations can be assessed and certification gained.
  ▷ **Part 2:** Guide to implementing BS 11000-1 is a guidance document to support organisations seeking to implement a framework specification structured in terms of 'what', 'why' and 'how'.

The overall aim of BS 11000 is to help organisations establish collaborative working that can result in increased efficiency and transparency, leading to better cost and risk management, as well as levels of innovation not normally achieved in typical client-supplier relationships.

▶ UK Infrastructure Route map 2014

- Establishes a tool kit in which the principal components:
  ▷ Enable the adoption of the common characteristics and behaviours associated with successful infrastructure project and programme delivery, including:

  - Early visibility and commitment to the pipeline of programme opportunities or the specific project.

  - Clearly articulated sponsor requirements adopting whole life principles linked to service outcomes that define the project or programme requirement.

  - Effective governance, accountability and timely decision-making.

  - Early supplier engagement that involves all tiers of the supply chain.

  - Effective use and structuring of standard contracts such as the NEC suite to align risk, reward and behaviours in an integrated supply chain.

  - Appropriate incentive approaches that stimulate further integration of the supply chain.

  - An environment that encourages innovation and departures from standards that embed cost and add no value to the outcome or safety.

- Provides a suite of assessment tools to enable sponsors, clients and the supply chain to align behaviours and identify capability gaps.
- Describes pragmatic approaches to compliance with EU procurement legislation.
- Offers an ongoing role for industry leaders and experts in the infrastructure sector to identify, develop and disseminate best practice.

▶ Alliance Agreements

Alliancing is a relationship between two or more parties which have aligned commercial interests and aim to work together to deliver a project in a collaborative and constructive way. An alliance agreement is no more than a consensual commitment between parties to work towards a common objective in a collaborative way. Such agreements have no contractual effect and are usually referred to as 'charters'.

Competitive selection procedures

When selecting an appropriate competition procedure it is crucial to know whether the client organisation is designated as being either public or private sector. For some projects that relate to public services it is also necessary to understand the structure of the client organisation and whether specific control and functional tests, first established by the Teckal[6] case, impact on this designation. A public sector body must comply (with certain exceptions) with the EU regulations which ensure that public sector bodies award contracts for projects that are above the specified value thresholds only after fair competition and only on the basis of the lowest price or the most economically advantageous offer.

When a contract does not fall into any of the exemptions provided for in the regulations, it is then necessary to establish which category it falls under. The covered categories as identified in the regulations are works, services and supplies. The regulations will apply in full to these various categories. However some services (Part B services) are covered by a special, less regulated regime.

A public sector body is free to choose a competitive selection procedure for projects that are under specified value thresholds for works, services and goods. When specified value thresholds are breached, a public sector body must adhere to the stipulated competitive selection procedures contained in the EU Regulations that vary between open, restricted, competitive dialogue and negotiated procedures.

Private sector client organisations are not bound at all by the regulations. However, BS EN ISO 10845: Construction procurement ISO 10845 does set out the steps and documentation entailed in a fair and competent tendering process. The underlying principles require the process to be:

▶ Fair: impartial and providing simultaneous and timely information, not prejudicing the interests of the parties.

▶ Equitable: non-award to a compliant bidder only if there are restrictions from doing business, incapability or incapacity, legality, conflicts of interest.

Transparent: procurement processes and criteria for each project/programme to be publicised and verifiable, with decisions publicly available with reasons given.

Competitive: system provides for appropriate competition to ensure cost-effective and best value outcomes.

Cost-effective: processes standardised with flexibility to attain best value in respect of quality, timing and price.

Public sector procurement routes

The following table (Table 4.2) summarises the four available competitive selection processes which are permitted for use in accordance with Directive 2004/18/EC (public sector). It is worth noting that Competitive Dialogue is not permitted where Directive 2004/17/EC (utilities) applies.

Table 4.2: The four available competitive selection processes for public sector procurement

Procedure	Summary
Open	This procedure is often used for the procurement of requirements which do not require a complex tender process.
	No negotiation with the bidders is permitted and there are no restrictions under the regulations as to when the procedure can be used.
	Under this procedure any potential contractor can submit a tender in response to the contract notice. Potential contractors may be instructed to provide a response to a request for information[7] (RFI) as part of a shortlisting or 'selection' exercise and they will also submit a response for quotation[8] (RFQ) tender return at the same time. However, this does not necessarily mean that every potential contractor's tender will be evaluated. The client organisation can evaluate all tenders if it wants to do so or it can decide to only evaluate the tenders from only those potential contractors who have met any prescribed selection criteria that may have been set (provided this stipulation is stated in the contract notice).
Restricted	Any potential contractor may express an interest in accordance with the contract notice and receive an RFI. Potential contractors who are successful at the RFI stage will then receive a RFQ and shall be invited to submit a further tender submission.
	No negotiation with the bidders is permitted and there are no restrictions under the regulations as to when the procedure can be used.

Competitive dialogue	Any potential contractor may express an interest in accordance with the contract notice and receive an RFI. Potential contractors who are successful at the RFI stage will then receive an Invitation to Submit Outline Statements (ISOS).
	The potential contractors shortlisted from RFI will enter in dialogue with the client organisation and individually discuss all aspects of the contract. Solutions are worked up with each potential contactor on the basis of the ideas and proposals put forward by that bidder. There can be no 'cherrypicking' by the client organisation of the best bits of the various potential contractors solutions, except with the consent of those concerned. Once the dialogue has generated potential solutions that meet the client organisation, all remaining potential contractors receive an formal invitation to submit final tender (ISFT) based on their individual solutions. The best potential contractor can then be selected, but there is very limited room for any further changes to be made once responses to the ISFT are submitted.
	This procedure can only be used in the limited circumstances described in the regulations.
Negotiated without prior advertising	There are two types of negotiated procedure which relate to whether or not a contract notice has been issued.
	Under the negotiated procedure without prior advertising, the client organisation is not required to issue a contract notice and may negotiate directly with the potential contractor of its choice. This variant of the negotiated procedure can only be used in the limited circumstances described in the Regulations.
Negotiated with prior advertising	Under the negotiated procedure with prior advertising, a contract notice must be published. All potential contractors may express an interest in tendering for the contract but only those meeting RFI selection criteria will actually be invited to do so.
	Potential contractors are invited to negotiate the terms of the advertised contract with the client organisation. The EU Procurement Regulations do not provide any rules to govern the conduct of negotiations, which means that departments can, within certain parameters, establish their own procedures for the negotiation and tender stage.
	This procedure can only be used in the very limited circumstances described in the regulations. In the event that the procedure leads to the award of a contract by a department using the open procedure, the restricted procedure or the competitive dialogue procedure would be discontinued because of irregular tenders or unacceptable tenders following an evaluation.
	Exceptionally, the nature of the work to be done, the goods to be purchased or hired, the services to be provided under the contract or the risks attaching to them, may not permit prior overall pricing.
	In the case of a public services contract, it may be that the nature of the services to be provided and, in particular, in the case of some specified services and intellectual services (for example, those involving designs) make it impractical to produce specifications with sufficient precision to permit the award of the contract using the open procedure or the restricted procedure.
	This procedure can be used in the case of a public works contract, when works are to be carried out solely for the purpose of research, testing or development but not with the aim of ensuring profitability or to recover research and development costs.

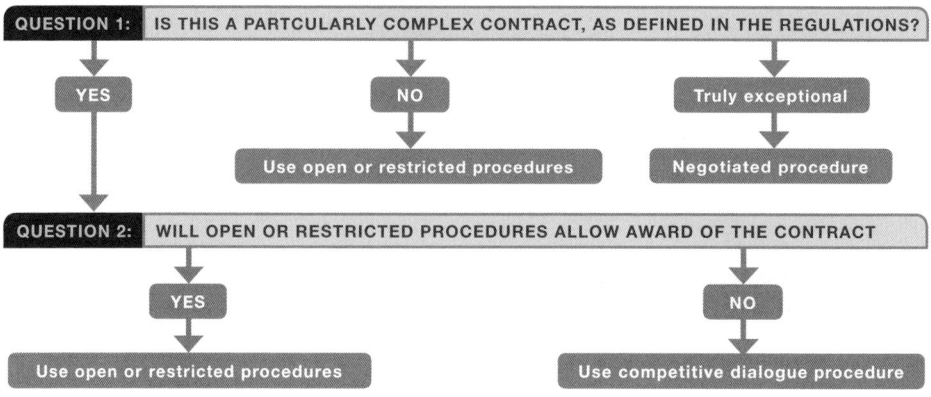

☐ Figure 4.1: The competitive selection process

The application of the competitive selection processes is illustrated in Figure 4.1 above.

How should clients get appropriate advice?

Procuring projects can be a complex process that requires the seamless bringing together of differing variables that were listed earlier in this chapter.

Whilst the reflex response to the challenges presented by a given project may be to consult a suitably qualified lawyer, there are a number of more important considerations including:

▶ Strategic project advice from an RIBA Accredited Client Adviser.

▶ Professionally produced designs and specifications from the right architect for the job.

▶ An accurate cost estimate from a cost consultant.

▶ Selecting the right contractor.

▶ The services of a project manager.

What are the basic features of each contract delivery method?

The characteristics of the delivery contract need to integrate with those procurement considerations that are defined at the start of this chapter, which are those of cost certainty, risk transfer, the need to accommodate change and the required level of quality.

A summary of contract delivery methods is below.

Turnkey or GMP

▶ Also known as 'GMP' (guaranteed maximum price).

▶ Usually based on bespoke terms (only one standard form – FIDIC Silver Book).

▶ Contractor takes most of the risk.

▶ Contractor usually deals directly with the employer.

See Figure 4.2 on facing page.

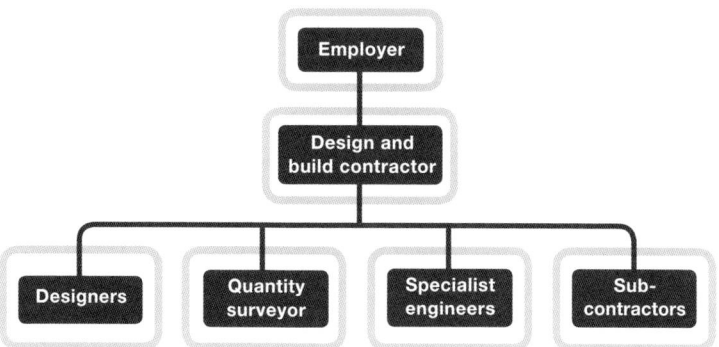

☐ Figure 4.2: Organogram for Turnkey or GMP procurement

D&C (Design and Construct)

▶ Can either be single Stage Two stage or Develop and Construct (hybrid two stage).

- In single stage the contractor provides a fully detailed offer based on comprehensive employer's requirements,

- For Two Stage the contractor is selected via an initial tender based on preliminary costs and programme, based on outline employer's requirements.

- Contractor takes specified risks.

See Figure 4.3, below.

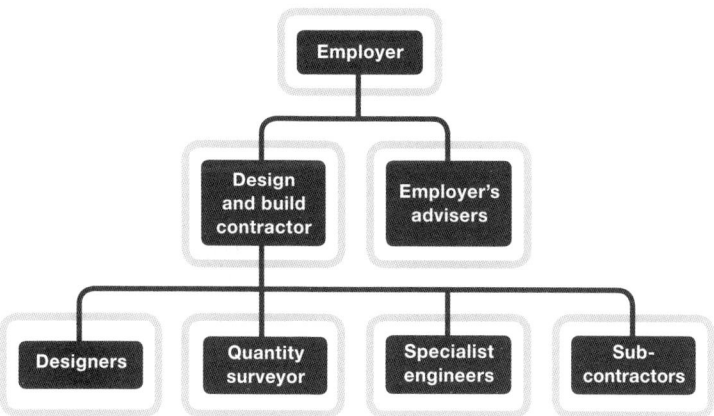

☐ Figure 4.3: Organogram for Design and Construct procurement

Develop and Construct

▶ Mechanism

- Design team initially appointed by the employer, in contemplation that it will be novated to the contractor after RIBA Stage C or D.
- The employer separately appoints project manager, employer's agent, cost manager and CDM coordinator.
- The design is completed to Stage C or D; meanwhile the contractor is selected and appointed on a pre-contract services basis.
- The contractor's initial appointment is typically made on a qualitative assessment (with or without the cost of preliminaries, overhead and profit for the works to be comprised in the Building Contract).
- The design team is usually (but not always) novated to the contractor upon its appointment.
- The contractor works with the design team and employer to develop the scheme in accordance with the brief and budget.
- Downstream procurement strategy is developed jointly by the contractor, employer and employer's advisers.
- The contractor may elect (or may be required) to involve key subcontractors in the design development and tendering processes.

▶ Advantages

- Effective management underpinning a fair allocation of design, performance and construction risk.
- Early validation of strategic programme and construction logistics.
- Full integration of design and construction through collaborative working.
- Overlapping of design and procurement without risk of unpriced design development.
- Reduced need for an additional shadow design team, where the original team is novated.
- Progressive coordination of the work of specialist contractors.
- Risk of information release lies with the contractor.
- Option for contractor to engage employer's architect and engineers directly.
- Risk of design detailing rests with the contractor.
- High scope for value engineering.
- Low employer risk to cost and programme overruns.
- Single point responsibilities with contractor.
- Shorter tender documentation period.
- Capable of conversion to guaranteed maximum price.
- Contractor buy-in to cost and programme.
- Early contractor input on programme, construction methodology and content of subcontract packages.

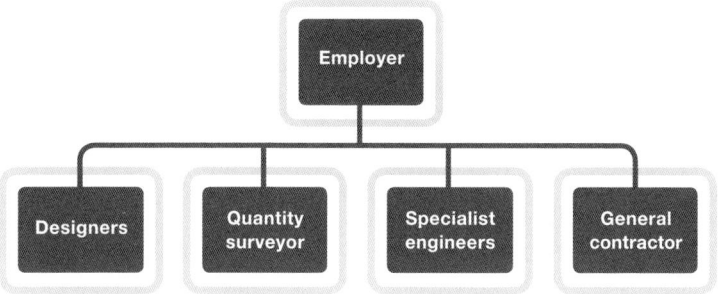

☐ Figure 4.4: Organogram for Fixed Price procurement

▶ Disadvantages

- May attract cost premium for post-contract design development and risk.
- Quality needs to be carefully defined and supervised.
- Post-contract changes unlikely to be uneconomic.
- More difficult for employer to control choice of subcontractors.

Traditional fixed price

▶ The traditional form of contract is based on a fixed price (lump sum) derived from full drawings, a detailed specification and either bills of quantities or an activity schedule.

▶ The work is subject to re-measurement if the scope varies.

See Figure 4.4, above.

▶ Mechanism

- Competitive tenders are based on complete design drawings for all elements.
- Measured bills of quantities or, alternatively, activity schedules are part of the tender documentation package.
- Detailed preliminaries are provided.
- Some elements may be the subject of Contractor Design Portion Supplements (CDPS) within the contract for specialist products and services for example cladding.

▶ Advantages

- Greater cost certainty provided; total cost known at outset of contract (providing scope remains unchanged).
- Better cost control of variations.

- Programme certainty – time frame established at outset of contract.
- Design responsibility remains with client.

▶ Disadvantages

- Relies on full design information prior to tender.
- Longer overall pre-commencement period required to develop complete design prior to seeking competitive tenders.
- Full design not usually achievable; leading to some Contractor Design Portion Supplements.
- Client is penalised if major changes introduced during contract.
- Cost risks remain with the client.

Target cost (may be used for traditional or D&B)

▶ Mechanism

- A target cost for carrying out the works is agreed between the employer and contractor.
- The contractor is paid the actual cost incurred (less some disallowed costs) up to the agreed target cost.
- If the final cost of the works is less than the target cost, the contractor and employer each benefit from receiving a pre-agreed percentage of the shortfall.
- If the final cost of the works exceeds the target cost, the contractor is liable for a pre-agreed percentage of the overspend.

Managed/management contracting

▶ Mechanism

- Design team appointed and retained by the employer throughout the project.
- Shortlisted contractors are invited to bid on the basis of their fees, method statements and preliminaries based on available information and the requirements of the strategic programme.
- Bidding contractors are interviewed formally following the submission of their bids to discover their corporate and personal credentials, to ensure that the successful bidder will fit well with the employer's team.
- A management agreement usually (but not always) consists of two parts: pre-construction and construction.
- The management contractor subcontracts construction works as a series of works contract 'packages'.

▶ Advantages

- Rapid method of procuring a contractor.
- Enables an early start on-site.
- Early validation of strategic programme and construction logistics.
- Facilitates 'fast-track' construction.
- Well-suited to large complex projects.
- Flexibility – design can extend into the construction period.
- Flexibility – later procurement of finishes to suit the design programme.
- Easier incorporation of changes into the contract.

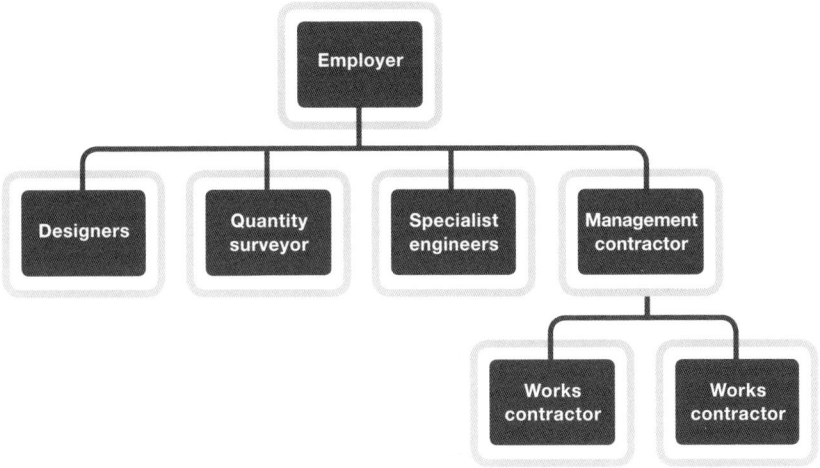

☐ Figure 4.5: Organogram for Management Contracting procurement

- Early contractor input on programme, buildability and the contents of subcontracted packages.
- Programme and cost plan agreed with the design team before the work starts.
- Engages the employer in rolling decision-making throughout the construction period.
- The content of works contract packages can be precisely detailed prior to procurement.
- In practice, subcontractors take more ownership of programme and quality issues.

- Useful employer participation in the choice of subcontractors.

▶ Disadvantages
  - Minimal risk for the contractor.
  - Close attention is required to optimise the interfaces between the subcontracted packages.
  - The subcontract strategy may not always align with the design method.

See Figure 4.5, above.

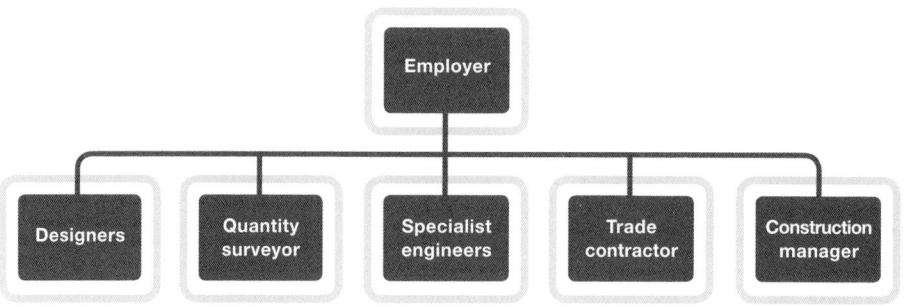

☐ Figure 4.6: Organogram for Construction Management procurement

Managed/construction management

▶ Mechanism
  • Procurement of construction manager as for management contractor.
  • Work packages are arranged by the construction manager as discrete trade contracts between the employer and trade contractors.

▶ Advantages
  • Rapid method of procurement.
  • Enables an early start on-site.
  • Flexibility – design can extend into the construction period.
  • Flexibility – later procurement of finishes to suit the design programme.
  • Easy incorporation of changes into contracts.
  • Benefits from the early input (from the construction manager) on programme, buildability and the contents of trade contract packages.
  • Programmes and cost plans are agreed with the design team, including

information release dates before the work starts.
  • Trade contract content can be precisely detailed prior to procurement to minimise works which fall between packages.
  • In practice, trade contractors take greater responsibility for programme and quality issues.
  • Creates the ethos of the 'single commercial team'.
  • Provides greater employer input into the choice of trade contractors, which requires a higher level of employer involvement in making decisions throughout the programme.
  • A direct relationship between the employer and the trade contractors may be attractive to the latter and can lead to their improved performance.
  • The employer retains the ability to influence the direction of the project.

▶ Disadvantages
  • The construction act places the employer at risk of direct adjudication from trade contractors.

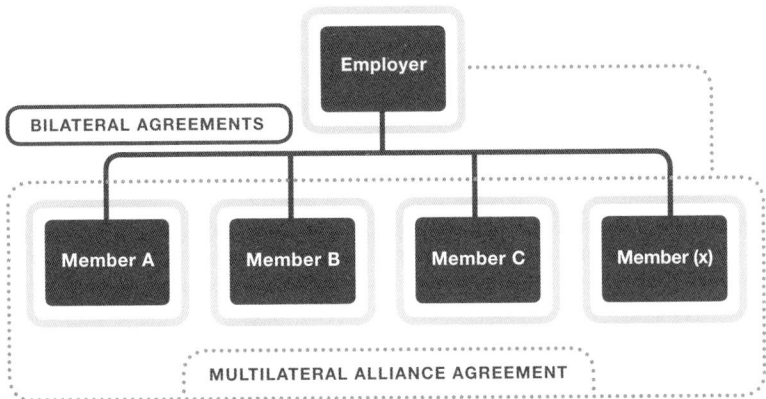

☐ Figure 4.7: Organogram for an alliance model for construction procurement

- There is a greater administrative burden to the employer who has to deal with numerous trade contract documents.
- Requires careful definition and scoping of roles and responsibilities.
- There is no construction manager buy-in to cost and programme – project delivery risk lies mainly with the employer.
- Approximately 15% of project costs are cost-fixed when the employer enters into the first trade contract.
- Close cost control and detailed reporting is required.
- Close control of procurement and construction programmes is required.
- There is a risk to the programme, enduring well into the construction phase as the scope of work develops.
- The employer is exposed to greater contractual risks.

See Figure 4.6, on facing page.

Alliance

There are a number of alliance models in current use, particularly within the infrastructure sector. Alliance models can differ in structure but the fundamental principles (as discussed above) are common to all.

▶ Mechanism (for the model shown in the diagram above):

- Operates as a series of separable bilateral agreements between the employer and each member.
- Alliance members are both individual contractors and professionals.
- Bilateral (commercial) agreements are based on a target cost contract mechanism (see above), where benefits and liabilities are transferred into and out of a central sinking fund held by the employer.
- A multilateral alliance agreement binds the members together.
- Each member benefits from a pre-agreed percentage of the sinking fund.

See Figure 4.7, above.

PPP (public/private partnerships)

▶ PPPs can cover all types of collaboration between the public and private sectors to deliver policies, services and infrastructure.

▶ PPPs are arrangements typified by joint working between the public and private sectors.

▶ Where the delivery of public services involves private sector investment in infrastructure, the most common form of PPP is the private finance initiative (PFI).

▶ The perceived advantages of PPP include value for money, better risk allocation and reduction in whole life costs, while providing better incentives to perform.

▶ However, the downside is loss of public sector control by the public sector over budget, privatisation of public services and greater cost in the long term.

▶ Figures 4.8 and 4.9, on the facing page, summarise PPP possibilities and processes.

Notes

[1] HM Treasury, Improving Infrastructure Delivery: Project Initiation Route map Handbook Version 1.1., (2014)

[2] Public Contract Regulations 2006 (as amended).

[3] Please note that Directive 2004/18/EC (public sector) and Directive 2004/17/EC (utilities), amongst other directives, are to be repealed in the UK with effect from 18 April 2016 with new directives planned to come in to force from this date.

[4] Application thresholds for the Directive 2004/18/EC and 2004/17/EC can be found in Commission Regulations (EU) No 1336/2013 (as amended).

[5] Requirements should include proposed investment in training and in health and safety.

[6] Teckal Srl v. Commune di Viano Case C-107/98 [1999] ECR1-8121.

[7] Please note that an RFI is synonymous with pre-qualification questionnaire (PQQ).

[8] Please note that an RFQ is synonymous with invitation to tender (ITT).

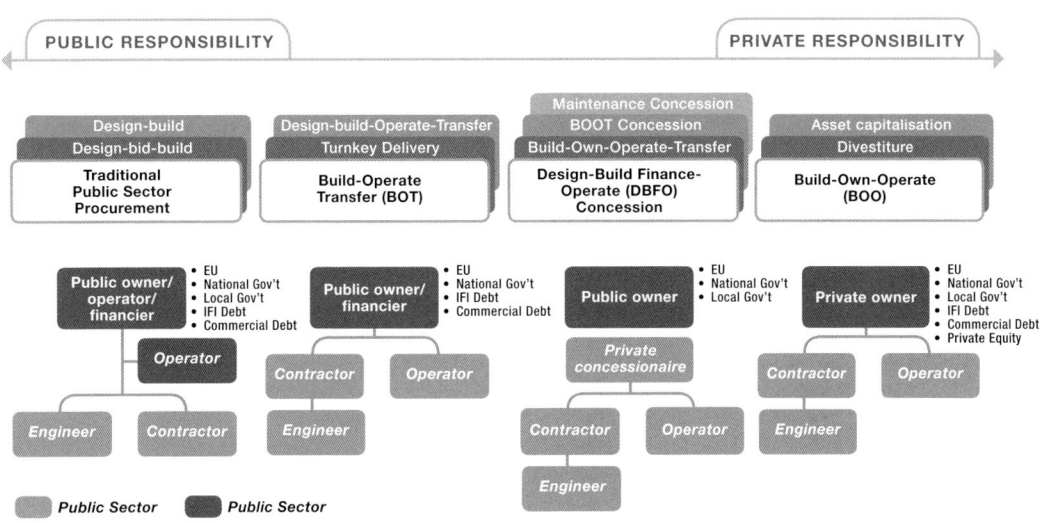

PUBLIC RESPONSIBILITY — PRIVATE RESPONSIBILITY

Design-build	Design-build-Operate-Transfer	Maintenance Concession	Asset capitalisation
Design-bid-build	Turnkey Delivery	BOOT Concession	Divestiture
		Build-Own-Operate-Transfer	
Traditional Public Sector Procurement	**Build-Operate Transfer (BOT)**	**Design-Build Finance-Operate (DBFO) Concession**	**Build-Own-Operate (BOO)**

Public owner/operator/financier
- EU
- National Gov't
- Local Gov't
- IFI Debt
- Commercial Debt

  Operator
  Engineer *Contractor*

Public owner/financier
- EU
- National Gov't
- IFI Debt
- Commercial Debt

  Contractor *Operator*
  Engineer

Public owner
- EU
- National Gov't
- Local Gov't

  Private concessionaire
  Contractor *Operator*
  Engineer

Private owner
- EU
- National Gov't
- Local Gov't
- IFI Debt
- Commercial Debt
- Private Equity

  Contractor *Operator*
  Engineer

Public Sector   *Public Sector*

☐ Figure 4.8: Public/private partnerships – allocation of responsibilities by type of arrangement

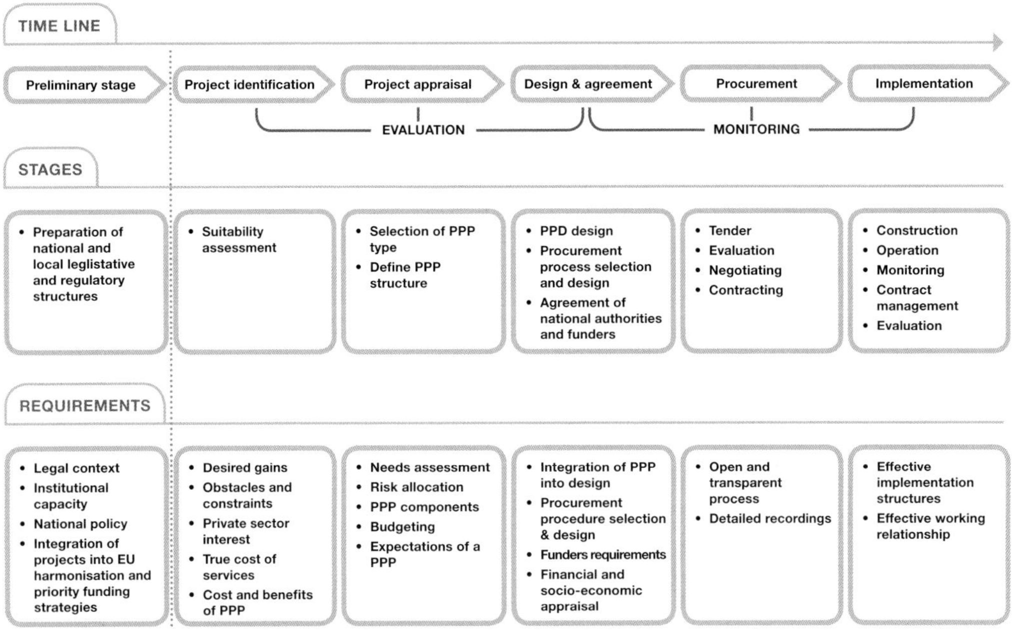

TIME LINE

Preliminary stage	Project identification	Project appraisal	Design & agreement	Procurement	Implementation

EVALUATION   MONITORING

STAGES

• Preparation of national and local legislative and regulatory structures	• Suitability assessment	• Selection of PPP type • Define PPP structure	• PPD design • Procurement process selection and design • Agreement of national authorities and funders	• Tender • Evaluation • Negotiating • Contracting	• Construction • Operation • Monitoring • Contract management • Evaluation

REQUIREMENTS

• Legal context • Institutional capacity • National policy • Integration of projects into EU harmonisation and priority funding strategies	• Desired gains • Obstacles and constraints • Private sector interest • True cost of services • Cost and benefits of PPP	• Needs assessment • Risk allocation • PPP components • Budgeting • Expectations of a PPP	• Integration of PPP into design • Procurement procedure selection & design • Funders requirements • Financial and socio-economic appraisal	• Open and transparent process • Detailed recordings	• Effective implementation structures • Effective working relationship

☐ Figure 4.9: Public/private partnerships – work stages and requirements

Construction procurement best practice

Gareth Hird in conversation with Peter Ullathorne

What are the main considerations in getting best procurement in the construction industry?

I think these can be summed up in the simple stages of:

- Have early conversations with all parties.
- Define the brief.
- Be realistic.
- Understand and take into account prevailing market conditions.
- Engage with constructors at an early stage in the design process.
- Appreciate everyone has the right to make a reasonable profit.
- Understand the relationship between time, cost and quality.

If you're doing all of those properly, you will be on course to succeed and should avoid the pitfalls which see poorly constructed projects delivered, or those that are late or over budget, or both.

Talking early on with potential suppliers is useful too, rather than just springing a procurement exercise on them without warning. You may find some of the key construction firms who would be most suitable for a project are busy working on a pipeline of work that stretches several years ahead. If you come to them at the last minute, they may be unwilling at that stage to embark on four to six weeks of lengthy pricing work in order to submit a tender. Whereas if they had advance knowledge of this work and its associated tendering, they could assess if this was a project they would be suitable for and keen on, allowing them to plan the resources for the subsequent tender process a few months hence.

How about defining the brief? How does that work?

Clients need to rely on the most appropriate professional advisers to engage with them at the earliest stage of brief-writing. Typically RIBA Accredited Client Advisers would be used in this role. They will investigate and question the clients to discover their detailed accommodation requirements related to achieving their missions and purposes. They will advise the client in his choice of architects and other consultants as to who can best realise the brief in physical terms.

What's your attitude towards profit?

Clients may make the mistake of trying too hard to drive down professional and contractor costs. Once clients accept the necessity of a reasonable profit for all project participants, they will not end up trying to screw down the costs. Driving down costs as the number one priority can create pressure that is passed down the supply chain via subcontractors which results in more intangible and unforeseen risks being introduced, and contributes to a dilution in quality. Taking the approach of 'reasonable profit' from the outset will mean that the correct costs of a project should be clear to everyone involved.

Why is it so important to understand and take into account prevailing market conditions?

The construction industry is influenced by outside market forces and effective clients should be aware (through the advice of their professionals which should include a RIBA Accredited (Client Adviser) of how these conditions will affect the availability and prices of particular contractors, the cost of materials and specialist services. The state of the general economy at any particular point in time may be different to that prevailing within the construction industry. Given good advice, effective clients are able to predict the impact of, for example, a market recovery that will restore construction companies to 'busy' status and increase the cost of their tenders and materials.

Conducting a sound tender exercise and securing the best possible contractor appointment is not something that should be left to chance. Effective clients should do all they can to avoid being at the mercy of the market or whoever decides to walk through your door looking for work. It may be a cliché, but construction procurement really is 'horses for courses', so clients need to make sure that only organisations of the appropriate size and experience compete in a tender process.

How does the public sector fare in this sector?

Often just getting onto a framework can be a major challenge. The existing procurement process means that big-name firms frequently get onto frameworks because they have the resources to engage in lengthy and costly tender and framework exercises. However, visibility of their layers of suppliers and subcontractors may not always be clear, and each sub-layer will have an associated cost, including a fee for administering the framework.

Chapter 5 What services do clients need from their design team?

by Adrian Dobson

"In selecting the design team, the aim is to find the best balance between technical expertise, management competence, costs, professional service and, of course, design quality. The importance of strong working relationships between the team members is also critical, as effective collaboration lies at the heart of all successful projects."

Introduction

This chapter describes the professional design and management services that a client needs to deliver successful building projects and the various ways in which a client can commission appropriate professional advisers.

Commissioning architecture

Whether you intend to construct a new building, or extend or modify existing facilities, it is important to understand your commissioning requirements, to assemble the right team of design and other consultants, and to accurately define the services you require.

The RIBA Plan of Work 2013 framework

The RIBA Plan of Work was thoroughly updated in 2013 and is the definitive model for the building design and construction process in the UK, and of influence internationally. It reflects the requirements of modern design and procurement approaches. The RIBA Plan of Work is unique in providing a clear, widely understood and recognised process map, which is flexible for use on projects of all sizes and with all procurement methods. Activities are arranged around a framework of eight numbered stages:

▶ Stage 0 – Strategic Definition
▶ Stage 1 – Preparation and Brief
▶ Stage 2 – Concept Design
▶ Stage 3 – Developed Designs
▶ Stage 4 – Technical Design
▶ Stage 5 – Construction

▶ Stage 6 – Handover and Close Out
▶ Stage 7 – In Use

The RIBA Plan of Work 2013 sets out, via a series of task bars, the activities and schedules of service that need to be undertaken at each stage, but does not specify who on the project team will undertake what services or the form that the stage outputs will take. In terms of design and project leadership activities, these matters are dealt with through the schedules of services, which form key components of the professional services contracts for the professional team. Construction activities to be carried out by the builder are defined through the building contract. In Chapter 6, below, Dale Sinclair describes the RIBA Plan of Work 2013 in detail.

The collaborative project team

Assembling an effective collaboration between of client, design team and construction team is recognised as essential for project success. The key roles and appointments that go towards creating the design team are:

▶ Client: the party commissioning the design and construction of a project. The client may be an individual or a company or organisation. It is important that, wherever possible, the same individual or group represents the client consistently throughout the project process. The effective fulfilment of the client role, especially in the early project stages in defining the project outcomes and design quality objectives, is paramount to project success.

- Project lead: the professional responsible for managing all aspects of the project and ensuring that the project is delivered in accordance with the project programme. A professional project manager or the architect would typically undertake this role, but where the client has substantial in-house construction expertise, this role may be provided directly by the client organisation.

- Lead designer: The design professional responsible for managing all aspects of the design, including the coordination of the design and the integration of specialist subcontractors' design. For building projects, the lead designer will most commonly be an architect, but a designer from an engineering discipline could also undertake this role depending on the nature of the works.

- Architect: the architect undertakes the architectural design of the project, including at the early project stages carrying out feasibility studies and brief development. The architect may also act as project lead, lead designer and/or contract administrator.

- Construction lead: the party responsible for construction and the provision of construction advice in the early project stages. The contractor would be the construction lead from Stage 5 onwards, after the building contract has been placed, but another construction consultant might perform this role prior to work commencing on-site, if the procurement approach does not include early contractor engagement.

The range of other professional roles and appointments required will vary significantly depending on the size and complexity of the project. For a relatively simple domestic project the client may only need to appoint an architect to lead the project and coordinate design, and a structural engineer for any structural calculations and design required. More complicated projects are likely to involve a number of other professional services. Some of the other roles would typically include:

- Civil/structural engineer: the professional engineer responsible for carrying out the structural/civil engineering design.

- Building services engineer: the professional engineer responsible for carrying out the design of building services.

- Cost consultant: the professional responsible for producing cost information as the design progresses. This information may include the project budget, estimates of the construction cost and life cycle cost analysis.

- Contract administrator: responsible for administration of the building contract, including the issue of instructions and certificates

- Health and safety adviser: a qualified professional able to advise on health and safety aspects, as defined by legislation and health and safety best practice. The Construction (Design and Management) Regulations 2015 cover the management of health, safety and welfare when carrying out construction projects. Under these regulations, one member of the design team must be delegated and appointed as the 'principal designer' to plan, manage, monitor and coordinate health and safety in the pre-construction phase of the project. This includes identifying, eliminating or controlling foreseeable risks, ensuring the other designers carry out their duties under the regulations, and preparing and providing relevant information to the principal contractor to help them plan, manage, monitor and coordinate health and safety in the construction phase. The lead designer or architect will be the natural choice to undertake this role on most building projects, since the 'principal designer' should have meaningful responsibility for coordination of the design of the project. On more complex projects, a specialist health and safety adviser, as defined above, may be appointed to advise and assist the principal designer in discharging their duties.

In addition, there may be the need for other specialists to be appointed, such as landscape architects, sustainability consultants, planning consultants, fire engineering consultants, lighting designers, interior designers, acoustic consultants and party wall surveyors.

Some projects may require designers with special experience and expertise. For example, when the project involves an element of masterplanning, the client may need designers with urban design skills. Works to historic buildings may require the expert services of a conservation architect and the RIBA maintains a register of architects with specialist accreditation for historic building conservation, repair and

maintenance, details of which can be found in the 'Find an Architect' section at www.architecture.com.

The RIBA Accredited Client Adviser: Strategic client leadership

Many clients will have significant in-house expertise and resources to assist with developing the brief and assembling the project team, but others may need professional support and advice in establishing the project. If the client is beginning a complex project and does not have knowledge and experience in procuring buildings, then having a professional RIBA Client Adviser on board – independent of the design and construction team – to provide strategic advice and construction industry knowledge from the earliest stages, can help bring certainty and control to the whole enterprise, and safeguard your investment. On smaller, less complex projects, the project lead, lead designer or architect may provide this early stage strategic support.

A RIBA Accredited Client Adviser can work with you to help define and deliver the best long-term solution for your organisation, one that will fulfil the aims of the project, by providing services such as:

▶ Setting collaborative project outcomes.

▶ Advising on strategic decision-making, including masterplanning and property portfolio management.

▶ Facilitating stakeholder consultation.

▶ Developing the brief.

▶ Assisting with budget setting.

▶ Undertaking feasibility studies and options appraisals, including new build, refurbishment and extension alternatives.

▶ Advising on procurement options.

▶ Assisting with selecting and appointing the professional team.

▶ Giving early advice on sustainability and information strategies.

▶ Undertaking appraisals of design proposals.

RIBA Accredited Client Advisers can help the client achieve high quality and best value in the building project. The most important decisions are the ones made at the very inception, before the planning and design stages and long before any bricks are laid. It is at this early stage that the success – or otherwise – of the clients' projects are decided. More complex projects often have multiple stakeholders and users who all have an interest in the project outcomes and performance. Often the development of an effective stakeholder management plan, which will serve throughout the project cycle, is an important deliverable by the RIBA Accredited Client Adviser that can make a significant contribution to project success. RIBA Accredited Client Adviser services are generally concentrated in the early project stages (Stages 0–2), but sometimes they are retained through the later project stages to offer continuing independent advice and design review.

The RIBA offers a referral service for RIBA Accredited Client Advisers, who they select for their all-round procurement expertise, design experience, business knowledge and track record of delivering results in construction projects. The RIBA evaluates and accredits all of its RIBA Accredited Client Advisers on an annual basis. More information can be found in the 'Find an Architect' section at www.architecture.com.

Services at Stage 0

The early project stages are critical in ensuring that clients get the best value from their design teams and the right project outcomes; changes made later in the process will result in additional expense and delay. In Stage 0 (Strategic Definition), the client and their professional advisers will establish and test the business case for the project and develop the high-level strategic brief. The strategic brief may require a review of a number of alternative sites or options, such as extensions, refurbishment or new build.

Services at Stage 1

At Stage 1 (Preparation and Brief) the design team will be assembled and they will develop and refine the project brief and programme. It is at this stage that you will need to confirm the core design team members and put in place professional services, contracts and associated schedules of services. On a simple project the design team might only comprise an architect and structural engineer but for a more complex scheme there could be a large number of different designers and specialist consultants. Before Stage 2 (Concept Design) can commence, it is essential that the project team is properly appointed.

Methods of selecting professional consultants

There are a wide variety of approaches to selecting architects and professional consultants for the design team. Clients may wish to identify a shortlist of potential practices based on recommendations or whose work they already know and admire. Professional services firms maintain comprehensive websites on which clients can find out about services they offer, the key personnel in their team and their business culture, as well as seeing examples of their previous work. Many professional institutes offer referral services to help clients to identify potential firms. RIBA Client Services, for example, help clients create a tailored shortlist of architects with the appropriate skills and experience. They will listen to the clients' requirements and then recommend accredited RIBA chartered practices. The RIBA also publishes a number of client guidance documents that are available in the 'Find an Architect' section at www.architecture.com: these include *Client Conversations*, *Working with an Architect for Your Home* and *A Client's Guide to Engaging an Architect*.

Clients may decide that a more formal competitive selection process could best meet their project needs, in the form of an open competition, invited competition or, more commonly, competitive interview. RIBA Competitions has extensive experience of delivering high profile selection processes, run as architectural competitions or other competitive selection processes by its experienced team.

Projects in the public sector above certain minimum financial thresholds are subject to EU procurement legislation, and professional services for such projects must be procured in accordance with one of a range of specified selection and tendering procedures. Comprehensive information on EU thresholds and procurement procedures can be found at: www.ec.europa.eu/internalmarket/publicprocurement.

Questions to ask your potential design team

The client should interview each of its shortlisted practices to gain a better sense of whether there is the right cultural and business fit to make the project a success. Building projects take a fairly long time from inception to completion and the relationship between the client and design team will need to meet the test of time. The interview should enable the client to test how compatible their culture, philosophy and personality are with their own working methods and its project requirements. It is a good idea to meet sufficient firms to be able to gauge and measure a range of alternatives without becoming swamped by too much choice and information.

The client will want to know about the experience the firm has built up from its portfolio of work and to check whether it has any necessary sector or specialist knowledge. Does the firm have a record of project delivery on time and within budget to required quality standards? What is their approach to the design process and do its completed projects reflect the sort of design aspirations the client has for its own project? The client should ask what the firm considers to be the most important issues and biggest challenges on its project.

The client may wish to visit previous projects undertaken by the practice, as well as taking up references from previous clients. This will give the client an indication of the quality of its work and perhaps also the opportunity to discover how other clients have approached projects with similar briefs, or consider design approaches in other sectors that might effectively cross over to the client's own programme.

Sustainability in all its aspects, ranging from energy and water efficiency to waste reduction and flood resilience, is a key element for a successful project outcome for many clients, and this is an area the client may wish to explore in greater detail with the potential members of its design team. If the client wishes to utilise specific sustainability measures, standards or certification on its project, such as BREEAM developed by the UK Building Research Establishment (BRE), LEED provided by the US Green Building Council or Passivhaus, the client will want to know about their knowledge and experience of these different design and assessment processes.

BIM

Building Information Modelling (BIM), using sophisticated computer-based 3D modelling with integrated specification, performance and cost data, is bringing about changes in the processes and working cultures of the building industry.

BIM can:

▶ Provide greater accuracy in design and construction planning and management.

▶ More efficient collaboration within the design team and between the designers and the builder and subcontractors.

▶ Richer data for project decision-making and post-completion facilities management.

Even if the client does not have specific requirements for BIM data, it may wish to understand the capabilities and resources that firms can offer to utilise BIM as a project tool that can improve project delivery and outcomes.

Viability

All the design and consultancy practices that make up the design team need to be able to provide adequate resources to support the client's project, and the client should seek clarity about who will lead the project for them and which staff will be working on it. It is also important to be satisfied about the financial resilience of the firms and to be certain that they will have the resources to complete the job, so company financial checks should be made before appointments are confirmed.

Fees

The client will also need information about fee levels and the payment arrangements. The fee is a matter for negotiation and agreement; there is no standard or recommended basis for fee calculation. Fees will reflect the complexity of the project, the scope of services to be provided, the work stages for which services are required, the procurement method (traditional, design and build, construction management) and the timeframe for completion.

Working relationships

In selecting the design team, the aim is to find the best balance between technical expertise, management competence, costs, professional service and, of course, design quality. The importance of strong working relationships between the team members is also critical, as effective collaboration lies at the heart of all successful projects. 'Selecting the Team' is a free guidance document produced by the Construction Industry Council (CIC) that can be downloaded at www.cic.org.uk (June 2005) and provides more detailed guidance on quality based selection of professional consultants.

The project roles table and contractual tree

On smaller projects, design roles are often delete comma, with the architect, for example, providing the services of project lead, lead designer and contract administrator. For bigger, more complex projects, the design service roles may be allocated to a much larger range of different consultants, and in these circumstances a project roles table and a contractual tree can be useful tools in managing the allocation of roles and services.

A crucial early decision involves the form of building procurement to be employed, as this determines the stage at which the building contractor will become involved in the project and take on some design responsibilities:

▶ Traditional procurement: the design team

produces full technical design information at Stage 4 (Technical Design) and the contractor is then appointed following a tendering process at the end of that stage. The contractor and/or specialist subcontractors may carry out some further design work during Stage 5 (Construction).

▶ Design and build procurement: tender documents are issued earlier, at either the end of Stage 2 (Concept Design) or Stage 3 (Developed Design), depending on whether a single- or two-stage tendering process is to be used. In this approach, the contractor will begin to offer input to the design process earlier in the project cycle.

▶ Contractor-led procurement: with this type of approach the contractor is appointed at the start of Stage 2 (Concept Design) and assumes primary responsibility for managing the design development. Contractor-led procurement is only usually employed on relatively complex projects, where the early engagement of the contractor's supply chain and contractor management of project risk is sometimes perceived to be advantageous.

The project roles table sets out in a simple tabular format which roles are required on the project at each work stage, and which consultant will undertake those roles at each stage. The contractual tree sets out in a hierarchical diagram the formal contractual relationships between the different design team members providing services to the project. The number of roles will depend on the specific needs of each project. Early scoping of these roles is essential to ensure adequate budgets are allocated for professional fees and to enable schedules of service to be prepared. Once all of the roles are identified, an analysis can be made of who might undertake them at each stage and if some specific roles are not needed at certain stages.

It is quite likely that one firm may be appointed to undertake a number of different roles, for example:

- An architectural practice acting as project lead, lead designer, architect and contract administrator.

- A construction management consultancy acting as project lead, cost consultant and health and safety adviser.

- An engineering consultancy fulfilling the structural and building services engineer roles.

There are also circumstances when a role might move from one firm to another at some stage, for example:

- An RIBA Accredited Client Adviser might also perform services as an architect at Stage 0 (Strategic Definition) and Stage 1 (Preparation and Brief) before passing this to another practice to develop the design at Stage 2 (Concept Design).

- In design and build arrangements, one architectural practice might provide architectural design services up until Stage 2 (Concept Design) or Stage 3 (Developed Design) before another architect employed by the contractor assumes architectural design responsibilities for the later project stages.

- Similarly, a contractor's specialist mechanical and electrical subcontractors might take over design responsibilities for building services engineering design once a design and build contractor has been appointed, and develop the initial design produced by the professional building services engineer.

For smaller projects, there is likely to be much greater aggregation of roles and design services into a smaller number of professional appointments, and passing of roles will be much less frequent. On smaller projects consultants will be providing a full service across all project stages. The RIBA has produced a template project roles table which can be downloaded free of charge as part of the RIBA Plan of Work Toolbox available at www.ribaplanofwork.com.

Setting up professional services contracts for the design team

Work should not be started by any member of the professional team until a professional services contract (sometimes referred to as an appointment agreement) is in place for the delivery of those services. As a minimum, each professional services contract should:

- Identify the parties to the appointment.

- Allocate and define limitations of responsibilities and liabilities.

- Define the scope of work (usually through a schedule of services mapped to the work stages).

- Confirm the legal framework (form of law, etc).

Confirm the fee, method of calculation and payment (normally set out in a fees schedule).

Set out methods of dispute resolution (typically mediation and adjudication.

Define provisions for the termination of the agreement.

The professional services contract is designed to help the parties avoid misunderstandings and disputes, to ensure that the fees and provisions for payment are clearly defined and enable the designers to grant a copyright licence to the employer.

On very complex projects it may be necessary to engage lawyers to prepare a bespoke agreement. Negotiations will typically focus on issues such as:

Arrangements for engagement of subconsultants.

Provisions in relation to deleterious materials.

Professional indemnity (PI) insurance provision and caps on liability.

Liability periods.

Novation of designers (in design and build arrangements).

Joint and several liability.

Liability periods.

Net contribution provisions.

In general, there is much to recommend the use of standard forms of professional services contract, and it is important for clients to remember that any significant change in balance in the apportionment of liability and risk to the consultant is likely to be reflected in the fees charged. It is important to ensure that any provisions agreed fall within the cover of the consultant's PI.

Services during the core design and construction stages – Stage 2 (Concept Design), Stage 3 (Developed Design), Stage 4 (Technical Design) and Stage 5 (Construction)

A design responsibility matrix can be used to describe the allocation of detailed design responsibilities for each element of the building and to ensure that the contractual schedules of service for each of the design team members accurately reflect these responsibilities. The schedules of service will also need to incorporate the requirements for administration of the building contract and site inspections, and reviews of progress.

Completing the project cycle

At Stage 6 (Handover and Close Out), the consultant team may provide services to facilitate the successful handover of the building, including updating 'as-constructed' information, certifying practical completion and the inspection of making good of defects.

Stage 7 (In Use) provides the opportunity for further services to be commissioned relating to gathering feedback on the project and assisting with the successful operation and use of the building. Post-occupancy evaluation of projects is seen as increasingly important by many clients, particularly when they are repeat clients with a portfolio of buildings and projects to manage.

The Building Services Research and Information Association (BSRIA) has developed a process, Soft Landings, designed to assist the construction industry and its clients deliver better buildings. Soft Landings aims to avoid a performance gap between design intentions and operational outcomes, for example:

▶ At Stage 1 (Preparations and Briefing), when expectations and requirements are set but may not be informed by experience and feedback from the outcomes of other projects.

▶ At the main design development Stages 2–3, when specific environmental and other performance targets are set and regulatory compliance achieved, but those targets may not be revisited or validated during Stage 4 (Technical Design) and Stage 5 (Construction).

▶ During Stage 5 (Construction), when budget shortfalls or value engineering may compromise the best of intentions, and variations to the building and its technical systems change how the building will be used.

▶ During Stage 6 (Handover and Close Out), when commissioning of building systems and user training may be rushed to meet deadlines.

▶ During initial occupation and use of the building, if not enough support is available to occupants and the managers to ensure the building is set up for the long term.

Getting the best from your professional team

Client Conversations is a RIBA publication that can be found in the 'Client Hub' subsection of the 'Find an Architect' section at www.architecture.com and provides insights into achieving successful project outcomes from the experiences of real clients in a range of sectors. It presents key lessons from a number of case study projects under six headings:

1 Defining project outcomes.

2 Leading from the start.

3 Assembling the project team.

4 Project briefing.

5 Mitigating risk.

6 Handover, use and feedback.

The contents will help the client to identify the services it needs from the design team and to get the very best from them in the delivery of the project. ■

Good client relationships

The ideal client is knowledgeable. Someone who knows his brief in advance, understands the design process, questions and challenges concepts, makes clear, timely decisions, appreciates value but is not penurious, is well funded and, above all, is fair in all his dealings with the design team. The ideal client does not exist.

A good client, however, usually has most of these characteristics, and with collaboration and guidance with his design team is drawn into a larger understanding of what the creation of architecture is truly about.

I have always believed that the best projects are the ones in which the client and team have engaged with one another from beginning to end. The client must realise that the job of his design team is to open and explore possibilities with him. At the same time, the client needs to be able to articulate his own particular vision and emphasis for his project, as well as state the problems he knows need to be solved – both the readily visible ones and the unseen issues that might come from organisational politics, company economics or professional preferences. Knowing your client and his objectives inside-out can save an immeasurable amount of time and effort during the design process.

Hopefully this knowledge will emerge at the beginning of the relationship, and as a corollary you will begin to understand how

best to communicate your ideas. Whether they are a universal quick study or a struggle to comprehend simple visualisations, each client is different in how they perceive your concepts. Your presentation may seem sensationally clear to you, but being certain that you understand how they understand is critical to moving process forward productively.

You also need to be certain that during the design process you truly understand what you are hearing from your client. Far too often we deceive ourselves with responses to leading questions we pose, looking for acceptance so we can move forward with our own agenda. Circling back and asking questions another way can help avoid the trap of 'hearing what we want to hear'.

Assuming you have a client who has done his homework, knows how to interpret a concept and can articulate critiques of the work, the next most important issue is that your client pays you fairly, on time and with a clear understanding of what he or she is paying for. You will learn much about your client as you negotiate your initial agreements. The burden is as much on you as him, however, when it comes to defining expectations for deliverables. At the end of the day, avoiding ageing accounts receivables depends upon all aspects of the client relationship but if you have got communications right from the start, they should never be an issue.

Chapter 6 The RIBA Plan of Work 2013

For organising projects by stages

by Dale Sinclair

"A core initiative of the RIBA Plan of Work 2013 is to provide the detailed support documents that define the specifics of who does what and when."

Introduction

The RIBA Plan of Work was devised over 50 years ago, evolving over the years to reflect changes in the industry. This chapter gives an insight into the purpose of each stage and task bar of the RIBA Plan of Work 2013 and how it might usefully be harnessed and used by effective clients.

The RIBA Plan of Work 2013

The 2013 plan is procurement agnostic although it is possible for an architect, project team, project manager or a client to produce their own practice or project RIBA Plan of Work 2013, which sets out more specific tasks for a number of common procurement routes.

It sets out the tasks undertaken by the project team, rather than the design team alone, and it incorporates a new 'In Use' stage (Stage 7), recognising the increasing importance of this stage and the benefits of using design data for in-use purposes or to improve a building performance or, more crucially, by assisting the development of future projects by linking Stage 7 to Stage 0. Post-occupancy evaluation studies would usually be a part of this stage.

There are a number of crucial points that need to be considered when using the RIBA Plan of Work 2013:

1   Although the stages can be used to define very precise tasks, the plan's limited word count cannot cover all of the detailed tasks undertaken on a project and it is intended solely to show the general thrust of activity at each project stage rather than the detail.

A core initiative of the RIBA Plan of Work 2013 is to provide the detailed support documents that define the specifics of who does what and when.

2   The plan sets out the tasks of the project team but does not allocate these to a particular party. As such, is not intended for use as a contractual document although the supporting tools assist in preparation of project specific multidisciplinary schedules of services and a design responsibility matrix for each member of the project team.

3   While it is available in a digital format, the plan is still widely used as an A3 pdf either as a generic template or a procurement-specific practice or project.

In summary, the RIBA Plan of Work 2013 is an essential project tool but the detailed and specific requirements dictated by different clients and the subtle nuances of each and every procurement strategy cannot possibly be captured on a single piece of paper. The plan therefore works alongside other tools that precisely set out, for every project, the detail of who does what and when.

What are the eight project stages and how do they impact on the client?

The eight project stages align with the UK government's suite of level 2 BIM (Building Information Modelling) documentation. The eight stages are:

Stage 0 – Strategic Definition

The purpose of Stage 0 is simple but strategic: it is to ensure that a building is required to meet

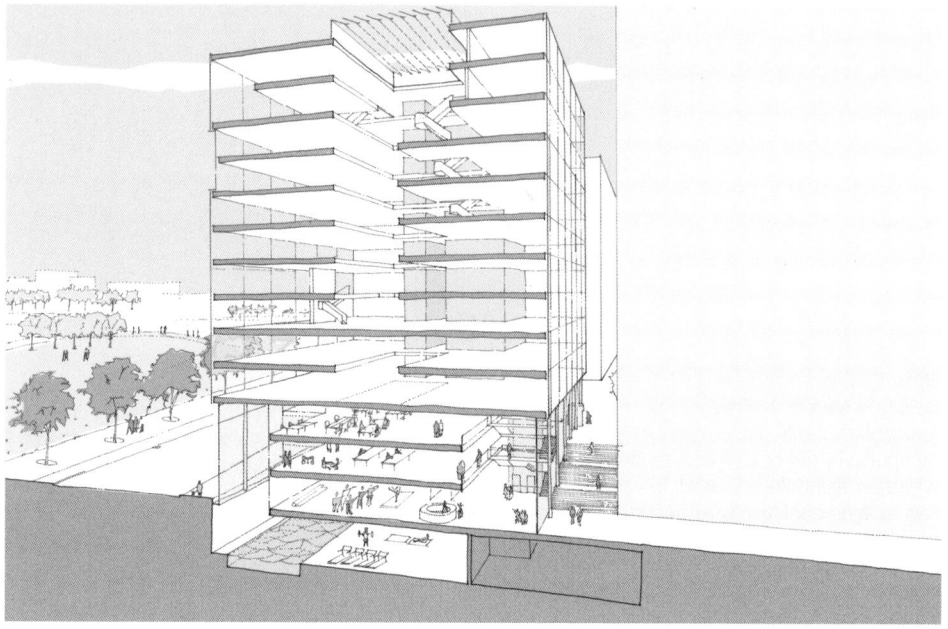

☐ Figure 6.1: Exploring the Concept

the needs set out in the client's Business Case and Strategic Brief – perhaps a refurbishment, space planning exercise or extension would meet the objectives in a more cost effective and sustainable manner? Clients may wish to consider appointing a RIBA Accredited Client Adviser at this early stage to provide them with important help, for example in the form of a route map, the services of a critical project friend and step-by-step guidance through the complex and demanding work within Stage 0 and all subsequent stages.

Stage 1 – Preparation and Brief

Having determined that a building project is required, Stage 1 puts more specific detail into the Strategic Brief with the generation of the Initial Project Brief. This brief will describe the Project Objectives in greater detail, including Quality Objectives, Project Outcomes, Sustainability Aspirations and other project parameters including the functional requirements and the Project Budget. Feasibility Studies may be required to confirm that the brief

can be accommodated on a given site. These tests can be essential and may flush out core briefing issues before design work begins in earnest. The project team will be assembled during Stage 1.

Stage 2 – Concept Design

During Stage 2, the Concept Design is prepared. The architect typically leads this process although the lead designer (who may be someone in the same practice or the same person on a smaller project) needs to ensure that strategic coordination issues are considered and embedded into the proposals. An RIBA Accredited Client Adviser would provide a range of useful services at this point.

The Concept Design can be presented and prepared using a number of different presentation tools. Many architects still use physical models to convey their proposals whereas others use sketches and freehand techniques, although the majority of practices now utilise various software packages to illustrate their proposals in 3D, including high quality visualisations (see Figure 6.1). Clients (particularly public sector clients) may require compulsorily use of BIM and, regardless of whatever technology is used, the client should require high quality information to be prepared by the design team for sign-off. BIM is described in Chapter 8.

Stage 3 – Developed Design

With the Concept Design signed off by the client, Stage 3 is fundamentally the lead designer's space. More detailed structural and mechanical and electrical exercises and analysis are undertaken, along with more stringent and detailed cost exercises. The Stage 3 Information Exchanges should comprise a coordinated design aligned to the Project Budget and validated against the Concept Design. While the design develops during this stage and, in some instances, certain aspects of the Concept Design may need to be revisited and discussed with the client, a fundamental reexamination of the Concept Design should not be undertaken. The outputs from this stage should be sufficiently robust to enable a planning application to be made. In some instances, a client may choose to use the Stage 2 outputs for this purpose, to keep cost down before the certainty of planning consent is achieved, although it needs to be remembered that the lack of validation that Stage 3 brings can present significant risks. Normally, it would only be experienced clients working with an established design team who would take this high-risk approach.

Stage 4 – Technical Design

The Technical Design stage delivers the detail required to build the project and consists of a mixture of design information produced by the design team and technical designs produced by specialist subcontractors appointed by the contractor.

The 'slicing and dicing' of Stage 4 will depend on the procurement route. On a contractor-led project, design work will be undertaken concurrently, whereas on a traditional project the design team will produce their Stage 4 information which forms the basis of the tender, with the successful contractor's specialist subcontractors producing their design work after the award of the Building Contract and in line with the Contractor Design Portion (CDP) clauses of the contract.

Stage 5 – Construction

Stage 5 consists solely of the construction process, including any prefabrication activities that may take place off-site. The role of the design team in Stage 5 depends on the form of construction contract used, with the extent of the team's involvement dependent on the procurement route and the needs of a specific client and/or contractor. The architect may or may not be the contract administrator and have on-site inspection duties.

Stage 6 – Handover and Close Out

Handing over a building has become increasingly complex due to the number of systems that are integrated into constructions. These might vary from a complex Building Management System (BMS) to a home cinema but either way the systems need to be commissioned and the users trained how to use them. Stage 6 acknowledges that the handover process can be complex, requiring consideration at Stage 1.

Stage 6 is currently framed around the traditional one-year period required to close out the Building Contract, tying-in with the initial three year aftercare period suggested in BSRIA's Soft Landings initiative.

Stage 7 – In Use

The RIBA Plan of Work has predominately been geared to the design and construction stages of a project, although the 2007 plan alluded to some In Use activities. The 2013 plan takes this a significant step further, sowing the seeds for significant tasks during this stage. These tasks are likely to be undertaken by new and different project team members but the crucial point is that Stage 7 will increasingly act as the gateway to a new project and Stage 7 to 0 activities will increasingly be part of framing the Strategic Brief.

The RIBA anticipates that a greater number of clients will harness this stage – particularly those responsible for building estates – to improve the effectiveness and efficiency of their future buildings by using Building Performance Evaluation (BPE) techniques that allow the analysis of their existing estates and for the data from such studies to form the basis of decisions for future estate management, including the briefing, design and construction of projects.

Do the eight plans of work task bars impact on the client?

The task bars and their purposes are as follows:

1 Core Objectives

The core objectives are indicative of the high level tasks that are undertaken during a stage. A Plan of Work glossary exists at www.ribaplanofwork.com/Help/Glossary.aspx and defines the capitalised and bold tasks in the plan of work in greater detail.

2 Procurement

The procurement task bar is generic in the plan of work template, becoming specific when a bespoke practice or project RIBA Plan of Work 2013 is generated.

3 Programme

The programme task bar has a single core function. It acknowledges that the general thrust of activity goes from Stage 0 to 7 but that on certain forms of procurement stages tasks will overlap.

4 (Town) Planning

The RIBA Plan of Work 2013 contains planning as a default activity at the end of Stage 3 but allows submission of a Planning Application at Stage 2 in a project or practice plan, subject to the risks being understood and/or mitigated.

5 Suggested Key Support Tasks

The suggested key support tasks mainly cover the supporting Project Strategies and other core tasks that support the core objectives set out for each stage. They ensure that topics like sustainability, buildability, and health and safety are considered holistically as the design is progressively fixed through each project stage.

6   Sustainability Checkpoints

The Sustainability Aspirations of a project are set at Stage 1 as part of the briefing process and are a core objective. The Sustainability Strategy is produced by the design team in response to these aspirations, at Stages 2 to 4 inclusive, and the contractor may also contribute to the strategy, as well as driving its content, at Stage 5.

The Sustainability Checkpoints task bar has been created specifically for the online version of the RIBA Plan of Work 2013 where the specific checkpoint activities for each stage are set out.

7   Information Exchanges

The Information Exchange task bar does not contain a great deal of detail but is fundamental in driving cultural change around the information that is exchanged at each project stage.

8   UK Government Information Exchanges

The UK Government considers certain Information Exchanges (data transactions) to be between the design team and the contractor, and is primarily interested in the data included in the Contractor's Proposals and the Building Contract and then the data provided at project handover for incorporating into CAFM (computer aided facilities management) or asset management software systems to ensure the more effective running of the project and to assist other Stage 7 activities.

What can we expect in the future?

In many respects BIM is an unfortunate acronym that focuses on information management and/or modelling. Increasingly, the term is being used as a description for change wrapped around many subject matters. The reality is that the disruptive digital revolution has just touched the built environment sector and profound and exciting changes are likely in the next five years, made possible by developments in cloud technologies, big data, the Internet of Things and artificial intelligence.

The RIBA Plan of Work 2013 is fully enabled for this exciting new world. ∎

What are the key characteristics and qualities of successful clients?

Clients are, self-evidently, at the heart of the life of any professional but the spectrum of behaviours that they exhibit never cease to amaze, amuse or infuriate professional advisers. It is worth looking at three specific attributes of the advice or the advisers that could be used as metrics for that spectrum.

Adviser or processor?

The first aspect is whether the client uses their professional as an adviser or as an instrument of process. The more satisfying (for the professional) role is to be a wise and trusted adviser as opposed to an operator of a process. This, however, requires that the client firstly knows what they do not know (cue Rumsfeld quotations) and are more than comfortable to admit it. That, in itself, is frequently difficult for many organisations and individuals when there is potential discomfort (or competitive risk) in admitting lack of Mastery of The Universe. At least as important is the client's ability to listen to advice, question the logic behind it, understand that logic and ultimately discuss the advice openly and constructively. This process by itself can unnerve and challenge weaker advisers but in so doing can help the client become a more effective user of professional advice. It is worth remembering the anecdotal comment attributed to JP Morgan (when a steel-man rather than a banker), who is supposed to have said, 'I want a lawyer who will tell me how I can do what I want to do, not whether or not I can do it.'

Proactive or reactive?

The better clients will value proactive advice rather than advisers who wait to be instructed. A good adviser is always thinking about their clients – particularly those whom they value and want either to keep or attract – and will be looking and listening to events and news, and offering an interpretation of either for those clients as a friendly (while never fee-seeking) thought. The better clients will react constructively to such thoughts and respect the willingness of the adviser to put their client's interest above the natural inclination of compliance people whose instinct will generally be to suppress any such thoughts in the absence of a formal engagement.

Advice or body shop?

There is a great deal of work performed by 'consultants' that does not honestly qualify as the provision of robust, independent professional advice but is merely the outsourcing of a task for which the client may not, currently, have either sufficient or sufficiently competent staff to perform the task. This sort of outsourcing of what might generally be the reasonable activity of the client has a couple of unwelcome side effects: first, that the standing and respect of the 'adviser' in such circumstances is significantly diminished, since it is not as highly valued as in other cases. These are often the cases that create the pejorative comments about 'consultants' around Whitehall and other parts of government – the negativity reflecting often genuine concerns about the value for money of the process. The second consequence of these sorts of engagements is the relative deskilling (or failure to upgrade skills) over time within those firms for whom this type of work is their more common mode of operation.

Conclusions

Firstly, that clients generally receive the advice they deserve over time. The relationship between client and adviser is symbiotic. Most good professional advisers are part of businesses that are more concerned with about the medium term profitability of the business and less about the impact of any action on quarterly sales figures – indeed it is arguable that the best advisers cannot function effectively in such a short-term financial environment as the quality of the advice will almost inevitably suffer.

Secondly, that both advisers and clients can both become better and more effective in their relationships with each other by identifying which mode suits each the best and then focusing carefully on the maintenance of that role with each other.

Thirdly, that the relationship itself, outside the execution of formal engagements, should be nurtured and supported by each side of the relationship continuing to think about the other. Advisers will see market or technical developments emerging before the majority of their clients. A wise adviser will use their acute knowledge to help clients understand the impact of the development – not in endless newsletters, flyers or spam-fodder emails (few of which are ever genuinely read by the real client) but in quick phone calls, coffee catch-ups or the occasional instant message. This on its own can be an extremely effective way of improving the quality of the advice when delivered, in that the adviser and the client can then be far more aware of each other's skills, concerns, focus and challenges at the point that the work begins. The context can be a strong underpinning of the engagement itself.

Chapter 7 The design process

by Rab Bennetts

"It is hard to carry out analysis dispassionately without seeing the potential of design ideas and, sure enough, some analytical sketches seem to correspond with others, thereby suggesting a possible direction or a 'vital spark' that might ignite an architecturally interesting direction."

Introduction

The acclaimed Italian architect Renzo Piano once described the process of designing buildings as 'a balancing act between art and science', encapsulating the constant tension between creativity and objectivity. Too much flair and there is a risk of the building's performance being compromised; too much functionality and its emotional appeal will suffer.

Each individual architect or architectural practice will have their own way of achieving the right balance but they will all agree on one thing: design requires a methodical approach that invariably follows a sequence of stages from conceptual ideas to the execution of construction details. The RIBA's Plan of Work 2013 and the Soft Landings methodology, described in Chapters 5 and 6, provide a universally recognised structure to this journey but what is the design process like in practice and who is involved?

Analysis

For someone who has not been involved in creating a building before, it is tempting to imagine that the architect has a blinding flash of inspiration that sets the path for the subsequent years of hard graft but, in reality, most architects begin with the rather dry process of analysis, on the basis that it is far better to understand the problem before coming up with a solution. The alternative – ie shoehorning a project's accommodation into a preconceived architectural form – has resulted in failures of all kinds, from cost over-runs to long-term problems of functionality.

The analysis takes many forms but invariably starts with the client's brief, responding to such questions as:

▶ What are the sizes and characteristics of each room or element of accommodation?

▶ What is the physical relationship between them?

▶ Are there any special technical requirements, such as long spans, security or temperature?

▶ How many people use each space or are there peak activities at certain times?

▶ What deliveries are needed?

The list of questions is of necessity very long for larger projects but can be condensed into simple diagrams over a short but intense period of information-gathering and roundtable brainstorming. Whilst the architect leads this process and acts as editor, there will be many contributors, from the client themselves of course, to structural and services engineers, quantity surveyors and specialists who might advise on issues such as traffic generation or acoustics.

The outcome may appear like a series of abstract diagrams but their value is immense, as they establish the key requirements of the project's functionality regardless of the visual or spatial composition of the design itself, which has yet to come. If the client is considering several different sites prior to purchase, these studies will still be useful as they are not necessarily site-specific.

Nevertheless, the analysis must also go on to consider external issues that do relate to individual sites:

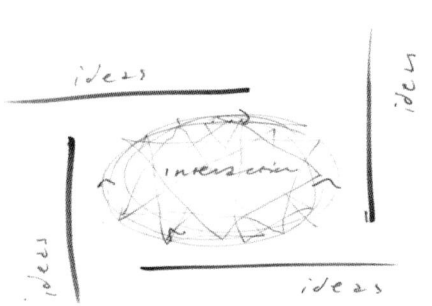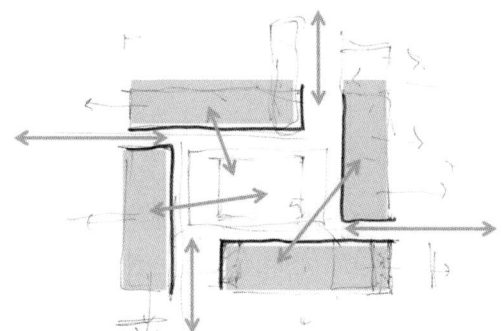

Figure 7.1 and 7.2: Exploring functional relationships at an academic building at Oxford University

- Are there any known planning policies that affect the site?
- Are there any party wall or rights-to-light issues?
- What are the restraints of topography or underground obstructions?
- Is there local traffic noise?
- Do sun angles or wind patterns affect the site?

Another series of diagrams, this time explicitly related to the site and its environmental context, start to identify issues to which any design must respond. Once again, the process is necessarily collaborative but directed by the architect as he or she will ultimately absorb the information into a design proposal.

This analytical stage is not entirely factual or unemotional, however, and it is important to foster a discussion about the project's purpose and to establish if there is something in its 'spirit' that might guide the formative stages of design itself. Take a library project, for example. With so many new forms of communication, why

would the public visit a library and what is the role of books? Is there something in the history of the site that suggests a building of stature or a public space as part of the project? Is there an atmosphere or a 'tone' that is appropriate? Should it recede into the background and allow the public domain to shine or is there a case for something assertive? Its future should also be debated, as a building tailor-made to its initial brief may be difficult to adapt in years to come.

Benchmarking comparable projects or visiting other relevant buildings together is valuable for agreeing common points of reference and for establishing team spirit.

Conceptual ideas

It is hard to carry out analysis dispassionately without seeing the potential of design ideas and, sure enough, some analytical sketches seem to correspond with others, thereby suggesting a possible direction or a 'vital spark' that might ignite an architecturally interesting direction.

Architects certainly do not have a monopoly on ideas and other designers – particularly

☐ Figure 7.3: Project design review

the engineers – may also contribute to early thoughts about shape and form, orientation or access, for example. Clients, too, should feel able to contribute by unloading ideas that they may have been thinking about for some time or by expressing opinions on the diagrammatic material being developed. Experience suggests that these first steps towards a design will prompt the architect to sketch out two or three options, some of which are exciting, others less so. Some architects prefer to debate their strengths and weaknesses with the team in order to set a clear direction, whereas others tend to lock themselves away for a few days to establish a single, well-worked-out proposal. Ideally, this formative stage should lead to a rough set of plans and sections, perhaps with a massing model to show how each option relates to its surroundings. It is probably too soon to know exactly what the building might look like but ideas about scale, materials or elevations may also start to emerge at this point. A really

bold client may consider the involvement of an artist, even at this early stage.

It is normally a thrilling stage for any project and the strength of the initial concept, backed by the client's support, will motivate the entire team for its duration.

Design competitions

Various types of competition are described elsewhere in this book but one cautionary note should be expressed here. Design competitions effectively compress the analysis into a very short period, often with little or no involvement by the client. Because of the arm's-length relationship between the client and the architect in a competition, many of the key conclusions of the analysis are a combination of architectural preference and guesswork. Whilst this may well produce an architectural proposal that is strong enough to win, it is also possible that a huge amount of work is wasted if the analysis has not had the benefit of client input.

Design development

As the design process moves from the general to the particular, initial conceptual ideas are increasingly subject to rigorous technical development as the structure and services principles are fleshed out and the form of construction becomes clear. Similarly, the architectural form develops from 'broad brush' to something that embraces the scheme's spatial potential, its colour or thematic details at a more tactile scale. Different architects approach design development in different ways of course, with some ensuring that the architectural aesthetic is utterly dominant whereas others will allow a degree of pragmatism to shape the final form. Suffice it to say that 'best practice' is represented by thorough integration of engineering and construction with the architectural form, such that it is economic, buildable and as free of known risks as possible.

However, it is worth noting that design is not a purely linear process and it is sometimes necessary to go back a step or two if the emerging design does not feel right or if an insurmountable problem has become apparent. During Stage 2 – the Concept Design – change is to be expected but during Stage 3 of the design process and beyond, only limited amendments can be countenanced without significant consequences.

Innovation and research

The design process sometimes requires some form of innovation, requiring research in order for the building to perform at its best. Perhaps the form of the building might cause wind turbulence, which needs desktop studies at least and possibly a wind-tunnel test in order to ensure it is safe for passersby. Sustainability and the pursuit of low-carbon buildings is another example where, say, new ventilation techniques need rigorous technical assessment before they can be implemented. New materials or methods of assembly inevitably require thorough investigation if the technical problems of the post-war period are not to be repeated.

All this takes time and expense and clients are entitled to ask for justification, but a limited degree of innovation is often necessary and is something any industry must undertake if it is to move forward. The research itself is unlikely to involve just the architect; external bodies such as BRE (formerly Building Research Establishment), specialist contractors or university departments are available on a case-by-case basis.

Whether or not to proceed with something innovative will depend on its impact on the project, its timing and, of course, a hard-headed assessment of its cost-benefit. In architectural terms, too, innovation can make a project distinctive, although there is a distinction to be made between scientific investigation and artistic originality.

Consultations

Third party consultation is now an accepted part of the design process and can determine a project's success. External criticism can be uncomfortable but, at its best, consultation can also inject interesting ideas that have not been aired before. By far the most influential consultee will be the local planning officer and the lesson here is to consult early and often; meeting just before the planning application is submitted is too late. The planning process has become increasingly complex over the last 20 years and it is now common for clients to engage specialist planning consultants to guide the design team through the bureaucracy and advise on consultations.

Much more difficult to gauge is the impact of consultation with local groups, some of whom will be genuinely interested in contrast to others who are resistant to change. Although it is not possible to please all of the people all of the time, genuine consultation (with a willingness to listen on both sides) is often productive and can strengthen a design through local knowledge and endorsement.

Design reviews are another type of external consultation, whereby a panel of experts carries out a critique of the design, normally at the behest of the local planning authority. For clients this can be an unsettling process as all kinds of comments are directed at the drawings or models on display but it is familiar territory for most architects through their university education. A credible, well thought through scheme will engender approval that greatly eases the route to planning consent but flawed schemes can expect a rougher ride. There are numerous bodies in the UK who offer design reviews of varying quality but the gold standard is held by Cabe (formerly the Commission for Architecture and the Built Environment), which is now part of the Design Council. The most productive form of review is an informal, confidential consultation prior to the planning

application stage but major schemes should also expect reviews to be on the record so they can form part of a planning officer's final recommendation to the planning committee.

Throughout these stages the architect can be expected to make frequent presentations, both to the client to explain the developing design and to outside bodies whose sanction may be needed. There is a real skill in communicating ideas at this critical point. Too little detail or too much can give the wrong impression; glossy presentations must be judged for their appropriateness against a couple of beguiling sketches; clarity of spoken or written language is far preferable to architectural jargon.

Team dynamics

By now it will be obvious that a successful design process requires the creative energy of a team that works well together. For most projects architecture is a collaborative process but the architect leading the team is still responsible for providing the overall sense of direction and coordinating the design emanating from engineers and others. In the later stages coordination becomes increasingly complex, as the overall design will have to incorporate the work of, for example, contractors who specialise in glazing systems or mechanical ventilation.

The client also has an important role creating the right conditions for good design – including the contract stages when some aspects of design are carried out by specialist contractors – empowering the team and being available to debate key decisions at the right time. Some clients can also find the design process stressful, as it examines their objectives in forensic detail, sometimes to the point of revealing that things do not quite add up: creating a building inevitably becomes a reflection of the individual or organisation who commissioned it. The best projects undoubtedly benefit in proportion to whatever time and energy the client can devote to it and the absence of a 'blame culture' can help to defuse problems at source. The relationship between the client and architect is therefore crucial, which is why so much care must be taken when selecting the team in the first place and why design competitions can fail simply due to the lack of good chemistry between the key individuals.

Sustainability

There can hardly be a better example of Renzo Piano's aphorism than the pursuit of sustainability through design, as the science is precise but the art is its physical expression. Sustainable buildings that no one likes are not only pointless but unnecessary, as much of the raw material for sustainable design supports a rich architectural vocabulary based on respect for the environment.

The purists would argue that true sustainability requires social, economic and environmental harmony but, in reality, this is hard to measure and there is no agreed standard to follow. By contrast, many of the solely environmental issues that apply to buildings are well documented – from water, waste and biodiversity to energy – and there is a regulatory framework that relates to long-term intergovernmental commitments on carbon dioxide emissions. The implications of CO_2 reduction in particular are profound, as buildings account for nearly 50% of the UK's emissions through day-to-day energy consumption and construction.

The science of CO_2 reduction is now embedded in the design process, with stage-by-stage assessments of performance as the design develops. A form of architecture that is based on 'passive' design is far more likely to meet CO_2 objectives than one that is reliant on mechanical systems that require significant energy to operate. For an office building, 'passive' design, in effect, reflects local climatic conditions, such as the use of thermal mass to stabilise temperatures, solar shading to avoid overheating, natural ventilation (where possible) to provide comfortable conditions, good day-lighting to reduce lighting energy, and so on.

In terms of architectural potential, then, a sustainable building should be more engaging than one that is not, thereby fusing science with art. With more than 20 years of 'sustainable' buildings now in operation, the broad conclusion is that simple and robust design strategies are best; undue complexity rarely endures.

As with architects who pioneer sustainable techniques, there are clients whose ambitions include being leaders in this field, sometimes for its own sake but also to reduce running costs or attract demanding young staff. Sustainability objectives should be debated as part of the initial brief and revisited as the project unfolds.

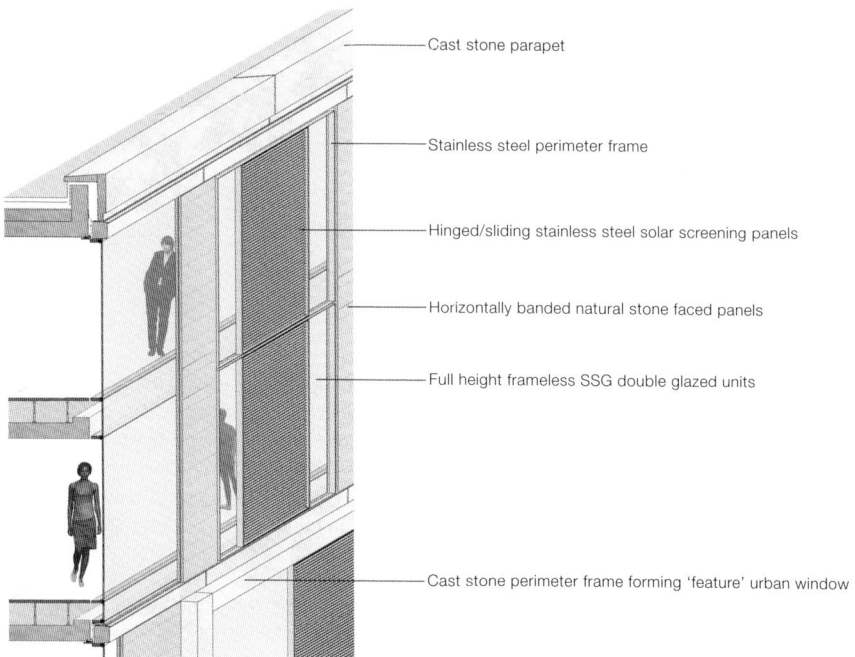

Cast stone parapet

Stainless steel perimeter frame

Hinged/sliding stainless steel solar screening panels

Horizontally banded natural stone faced panels

Full height frameless SSG double glazed units

Cast stone perimeter frame forming 'feature' urban window

☐ Figure 7.4: Exploring the envelope: function and appearance

Designing for construction

Although there are many pockets of world-class excellence in the UK, the design and construction industry as a whole has an unwelcome reputation for projects that are late or over budget. In the last 20 years or so there have been many attempts at industry reform, most of which have focused on integrating design and construct from an early stage, that are described on p59.

For the architect (and other designers such as engineers and landscape architects), this shift in emphasis amounts to a fundamental change in the status quo. No longer does the design team work exclusively for a client but it must also expect to be employed by contractors whose objectives are somewhat different, possibly switching partway through the same project. The skills required to reconcile these potential conflicts are quite different to those of the 'traditional' architect, and the balancing act is between quality and delivery, as opposed to

art and science. Furthermore, the construction industry itself now undertakes large elements of detailed design as part of its huge range of specialist trades, such as glazing systems, some types of roofing, mechanical and electrical systems, piled foundations and so on. Far from being the diminution of the design team's duties decried by some, this involvement in design by contractors can also be seen as a productive way of getting the best out of industry, in much the same way as the great architects of history worked closely with specialist stone masons or carpenters to achieve an overall result. The notion that the architect should know how to specify every detail of a complex construction is obsolete.

Nevertheless, it takes a particular temperament to switch from working directly for a client to a contractor and it follows that one of the most important design decisions is the choice of contractor. When attitudes are aligned and personal chemistries are working well,

Figure 7.5: The end of the process. St Antony's College Oxford

design and build is as good as other forms of procurement; by contrast, failure to achieve these may result in a state of attrition, with consequential compromises on quality.

There is also considerable skill in producing the right type of information at the right time; if it is too detailed, the contractor's price may be too high, whereas something less prescriptive allows the contractor to bring in specialists with their own proposals, always assuming they are compliant with basic qualitative criteria. When a precise detail really is important, it needs to be described 'up front' at the time of tender. With design and build, designers do not have the licence to enhance their ideas once the contract has been agreed, unless the client is prepared to risk additional costs.

Clients need not be alarmed at what might be perceived as delegating a limited amount of design responsibility to others but it is essential to understand and agree the limits to this approach before it is enshrined in a contract. Although architects and others still produce a substantial quantity of detailed design information for construction, much of their time in the latter stages is spent inspecting drawings prepared by others and providing coordination to ensure everything fits together.

As with the earlier design stages, the client can provide the right conditions for good design through adequate time and rigorous tender selection procedures. Avoiding changes is also a vital discipline if the project is to remain on track.

The end of the design process?

Completing all the design details might seem to be the end of the process but there is still much to be done. Site meetings, visits to suppliers' works and quality inspections take up a great deal of time and are as important to achieving the right outcome as producing good information. Good decisions made on-site in response to unexpected problems are design decisions of a type, so the designer needs to be there at the end as well as at the beginning, regardless of the type of contract.

Then there is the process of handing the building over when it is effectively complete, followed by final defects inspections and, in an ideal world, post-occupancy monitoring. Using the Soft Landings approach described elsewhere in this book (see page 5) could help greatly.

Unlike some other team members, such as project managers or quantity surveyors, the responsibilities of the architect, engineer and other key designers do not finish at project completion but last for many years so the long-term performance of the building must be considered alongside the shorter-term need for completion on time and to budget. To quote another definition of architecture, first coined by the Roman architect Vitruvius, the objective must be 'commodity, firmness and delight'; ie the building must fulfil its functions, be durable and please the senses. Having proved to be a robust definition for 2,000 years, the design process can be called a success if this trinity of aims have been achieved.

The planning system – *faites vos jeux*

Everyone knows that we have a planning system – but it is by no means self-evident what it is really for. In its origins in the first half of the 20th century, at a time of unregulated growth, it was devised principally to address serious concerns to do with public health and nuisance – for example, so that people did not have to live next door to a glue factory. But in less than 100 years, mission creep has set in and we now have a very complex system that can control – and in fact micro-manage – almost every aspect of building activity. Control of land use is still at the heart of the system, even though your neighbour who wants to set up a business is more likely to be using a laptop than boiling glue – but the system now concerns itself with every conceivable detail of amenity and environmental protection.

Each new planning minister says they want make the system simpler – each leaves office having made it more complicated.

The planning system, more than any other aspect of public administration, can be seen a paradigm of the 'nanny state'. But it has ended up like this in response to demand. Given the complexity of the system, it is no surprise that obtaining a planning consent, for a project of any size, usually involves a protracted negotiation – a process now at the heart of the system, although the word negotiation does not appear anywhere in the government's National Planning Policy Framework (NPPF).

The planning system tends to be more interested in how a proposal relates to the rest of the world than in the proposal itself – so a local authority's priorities may be upside down when compared with those of the client and their design team. And within a council, elected members may not have the same priorities as officers. Every significant planning application is looked upon as an opportunity for the local authority to get things that it wants – within the scheme itself, or outside it, via the 'Section 106 system' – a sort of legally sanctioned 'brown envelope' that allows councils to ask applicants to fund what the council wants to see happen, as long as some tenuous connection with the project can be demonstrated.

The planning system has to operate under the pretence that it is rational and objective, but as practised it is a black art, and highly political, both with a small and a big 'P'. A lot may depend on whether decision-makers like the look of your scheme – or indeed of you. And because there are so many different aspects that are negotiable, while any one point can be considered in the light of adopted policies and guidance, the degree of discretion within each subject area is such that in practice there is a table full of chips to be shuffled around.

Perspective

by **Peter Stewart**

Planning negotiations are generally carried out with the local authority's planning officers. Planning decisions for major projects, however, are made by elected politicians who will be thinking about what their electorate will think – even though their task is meant to be to decide whether a scheme complies with the planning policies that they have set out. Planning officers are professionals (unlike politicians) – but their advice to you may be tempered by attempts to second-guess what the elected members will think (and it could be in your interests that officers do this). The dynamic between officers and elected members can vary greatly from planning authority to planning authority. In some places, an authoritative or forceful chief planner may rule the roost, and members are content for that person to make the running and vote for or against schemes as they are recommended. In a neighbouring council, members may undermine officers at every turn.

Decisions can be capricious – officers may recommend a yes on the basis of 12 months of discussion, but councillors can say no in a matter of seconds at planning meeting, while you were trying to find the right page in the agenda.

The human factor is important in planning. A client wants their architect to design a great building for them. But without planning permission – in the UK at least – the great design will not get built. As well as being a designer, the architect needs to be a communicator who can convince officials and politicians of the merits of their design. They need to be able to tell a coherent story, in a persuasive way, about how the design is the right one for the place it will be built, as well as for the client's brief. They need to be able to listen to what is being said on the other side of the table as well, and to be able to judge when to stick, when to twist and when to fold.

In conclusion, my five top tips for clients who seek to find their way through the territory sketched out above are:

1 Personalities matter. Build relationships.

2 With your architect, present your case to the planners confidently, in a way that shows that you understand the planning system.

3 Remember that the planning system is more interested in the project from the outside in, while the client may be more interested in it from the inside out.

4 Work out what the council would like to gain from your project – which may be quite different from what you want from it.

5 As a last resort: remember you can appeal against a refusal. You will get a rational hearing from a planning inspector if you did not get one from the local authority.

Chapter 8 Intoducing BIM

by Richard Saxon CBE

"The arrival of BIM marks a switch for the construction industry from considering buildings as projects to considering them as assets. The circular model of the asset life cycle replaces the linear model of a project."

Introduction

Perhaps the biggest recent change for clients in working with the construction industry is the arrival of Building Information Modelling (BIM). BIM has been around by that name since 2003 and the government decided in 2011 that all its work – 20% of the total – will use so-called 'Level 2 BIM' by 2016. That is, pulling the whole industry to adopt the digital work method.

This chapter assumes that the reader is unfamiliar with BIM and answers the nine frequently asked questions:

1 What is BIM?
2 What is in it for clients like us?
3 Can you walk me through the BIM procedure?
4 How do we become BIM clients?
5 How do we appoint teams that can deliver BIM for us?
6 How do we value the costs and benefits of using BIM?
7 Can BIM help us with sustainability?
8 Can BIM help us with asset management too?
9 What comes next after Level 2 BIM?

This is a fast moving scene. Advice given in 2015 has the benefit of the relative completeness of the tool kit devised by the government's BIM Task Group. But further tools are in development and methods will evolve quickly under the pressure of application. New software offerings will open up new possibilities. Client best practice will continue to evolve.

1. What is BIM?

From a client point of view, BIM is the use of integrated digital technologies in the design, construction and operation of buildings and infrastructure.

Only the government is a big enough customer to pull through change. After much lobbying, the UK government came out strongly, taking an international lead in devising a way for today's available BIM to be used. A task group of industry experts has defined what is known as Level 2 BIM. Level 1 was the use, since the 1980s, of 2D and 3D CAD and document management.

Level 2 uses 'object modelling', where the computer recognises building elements and holds a database of geometry and other data about the element and the whole building. Each design profession continues to devise its own information model but comparison software enables the client and the project team leadership to view and coordinate the set. The models are 'federated' to reveal clashes between them and to run applications to rehearse the site assembly sequence and to derive costs. The client can require the final operation and maintenance information to be provided as digital data to use with their facility management software.

The earliest big BIM example in the UK was Terminal 5 at Heathrow airport, completed on time and budget and let down only by its errant baggage system. More recently, the Aon tower in the City of London – aka the Leadenhall Building or the Cheesegrater – demonstrates

BIM's ability to support prefabrication off-site. It was built quickly and economically, and without site errors. BIM is now being used successfully on projects below £20m in cost.

2. What is in it for clients like us?

For estate-holding clients, BIM offers not just the possibility of faster, cheaper and better quality construction but also a database of operation and maintenance information to support the life cycle. Clients who have preferred or required standards for their properties can hold these standards in a BIM library of model elements and provide them to design-build teams to incorporate into projects. Feedback can then add to or improve these standards.

Once the client and team are used to working in BIM mode, savings in time and cost should become apparent. BIM speeds up the process, provided that its required procedure is followed. Design decisions are concentrated into the early project stages while later production stages are partly automated. With the quantum leap in quality of information available to the builders, they can choose to use off-site prefabrication to speed site work; BIM can drive manufacturing tools. The builder can also organise the site more easily for efficiency and safety and schedule deliveries better; time is BIM's 4th dimension. Its 5th dimension is that cost information is a byproduct of the modelling process. The 6th dimension is the asset information needed to operate and maintain the building, available immediately on handover.

Cost is reduced mainly because time is saved and risk reduced.

Quality improvements flow from the potential to achieve zero construction defects resulting from errors in design information. Modelling can also enhance quality in the performance of the building by simulation of energy flows or user circulation. Most improvement can come from the inclusion of Soft Landings in UK BIM practice. This process was developed by the University of Cambridge and its consultants to make sure that its new buildings performed as planned. The operational requirements of the facility managers become part of the brief and Employer's Information Requirements. At commissioning stage, the operational pattern is trialled and debugged and the operators trained. After move-in, members of the design, contracting and subcontracting teams stay with the building for a proving period to make sure that all is working to plan. The typical 'performance gap' between design and delivered building is thus minimised. Government Soft Landings, the official version of the Cambridge scheme, adds user outcomes to the brief. In-use evaluation after two years' use measures the results against the business case for the project. The O&M (Operation and Maintenance) database provided by BIM is fundamental to the quality of building performance and a major source of whole-life economy. FM functionality in BIM is likely to increase substantially in the coming years.

3. Can you walk me through the BIM procedure?

This walk-through is based on the RIBA Plan of Work 2013, described by Dale Sinclair in Chapter 6, which was created to fit the use of BIM. It has eight stages to cover the project life cycle (Figure 8.1). The recommended procedure deals not only with information management but also with long-standing weaknesses in the construction process and industry behaviour.

Stage 0: Strategic Definition

This stage precedes the start of the project, so is called zero. The client, supported by their advisers, decides if there is a business case for the project, what outcomes are intended and what success will look like. Resources and site are identified and the client decides on how to buy the project from the industry, choosing a suitable procurement path. The principle of using BIM will be decided at this point, as will the principles of operation and maintenance, and their need for data and for the Soft Landings process. All this forms the Strategic Brief.

Stage 1: Preparation and Brief

Most of the client's BIM-related tasks fall here. The team must be formed to follow the chosen procurement path and the brief developed to allow design work to start once it is signed off. Use of BIM affects the appointment routine considerably. The client must set out their plan for the project with the scope of service required from each team member and the programme of information exchange and decision points.

The CIC BIM Protocol should be used to add BIM responsibilities to the typical appointment contracts of team members. The Digital Plan of Work (2015) should be used to define these responsibilities at each stage of the project. No other contract or insurance variations should be needed.

BIM requires the client to list their Employer's Information Requirements (EIRs) to guide the content of the BIM so that it answers their brief at each stage. They must be clear how they will assess the competence of bidders. The full checklist is set out in PAS 1192-2, the emerging standard for information management in BIM. Supporting guidance is found on the BIM Task Group site: www.bimtaskgroup.org. Depending on whether the contractor comes in from the start or after the concept design, procedures of building up the team will vary. An information manager, however, needs to be appointed – usually the lead designer or contractor. This role includes support and training of team members and the supply of a server called the common data environment (CDE), often a web-based service, to act as the repository for information exchange. Before the stage ends, the CDE should be running and the processes, skills and technology should be in place. The stage completes by carrying out the first of four Information Exchanges: the appointed team submit the outline brief for the clients approval, including the site model in BIM. This should support the client's decision to proceed to Concept Design.

Stage 2: Concept Design

At this stage the geometric outline of proposals appears in BIM. Clients usually request options to be offered and the end state of the stage will be the development of the selected option to show initial spatial, structural, services and cost information. Depending on your programme, the concept design may be tested for acceptance by the planners, but this will mean speculative visualisations ahead of true finality. Walk-through and set-piece visualisations and simulations are a feature of BIM-based design, and stakeholders and planners can appreciate the proposed answer to their brief more clearly. Changes caused by this testing allow the brief and concept to develop to approval economically, compared to any changes made later.

To work with the team in BIM, the client will be playing its part in a disciplined quality assurance method. Each consultant will pass its proposals for a part of the concept to the CDE 'team-shared-area' for approval by the team leader. It then passes to the 'client-shared-area' for approval and use outside. The client judges how it meets the brief and whether it satisfies the information requirements needed to make a decision. Automated tools can help with validation. The second Information Exchange covers the full concept and updated brief and cost, and releases the project to move to Stage 3.

Stage 3: Developed Design

This should be the final stage of deep client involvement before commitment to build. All internal issues should be cleared by this stage,

with all client questions satisfied. The contractor ought to join the team no later than this stage for good BIM performance so that buildability issues can be settled before final design. At Stage 3 the team selects the technologies they will use and the generic products required. These are represented in the BIM by 'placeholder' objects until actual products are selected. Planning and building control approval are reached at this stage. The Stage 3 BIM must answer all the criteria required to support a client decision to invest and build. This includes Information Exchange 3, fixing design, price and timetable.

Stages 4 and 5: Technical Design and Construction

The client has no need to be closely involved in these stages as they will be worked within the decisions taken at the end of Stage 3. The specialist contractors will be turning the generic Stage 3 BIM into specific information to build, choosing products that meet the specification and consulting with the original designers to complete their packages. The contractor will be overseeing manufacture and assembly on-site Information exchanges continue within the supply team but only break out if circumstances require a client-approved change.

Stage 6: Handover and Close Out

This new stage results from BIM ways of working. Rather than just being the end of construction, Stage 6 prepares the building and its data doppelganger for handover to the client. The client will have asked at Stage 1 for handover information to form the basis of an

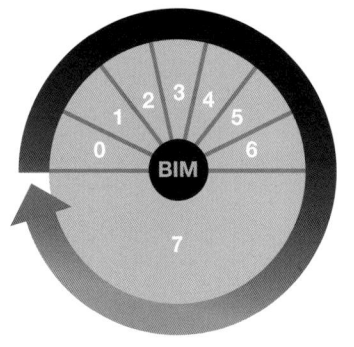

0	**Strategic definition**
1	**Preparation and brief**
2	**Concept design**
3	**Developed design**
4	**Technical design**
5	**Construction**
6	**Handover/close-out**
7	**In use**

Figure 8.1: RIBA plan of work 2013:
BIM Activity cycle

Asset Information Model (AIM), including the O&M database. Each component BIM will be handed over in native software format, together with a federated version of the whole. They will have been formatted for O&M purposes, with excessive detail removed to simplify the model. A set of pdf drawings can also be provided, for use by staff. Finally, the data in the BIM can be provided in spreadsheet format using a tool called COBie (Construction Operations Building information exchange). This US Army devised schema allows translation of the data in the model into the chosen computer aided FM programme for building and asset management. It is also a checking tool, validating content against requirements to save substantial manual labour. COBie should ideally be run throughout the project, from Stages 1 to 6, to check information exchanges at each level of detail reached.

Stage 7: In Use

With the building and its AIM now in use and Soft Landings running to ensure issues are ironed out, the project is in feedback mode. Project performance, including judging the effectiveness of the BIM working process, can be reviewed early on, before the team disperses. Have new standards been identified which should enter the client's BIM library? Have the standards used worked or could they be improved? Can the building's performance be improved further?

4. How do we become BIM clients?

Clients need to have a BIM Strategy. This starts with assessing the current level of maturity as a BIM user, identifies the level the client wishes to reach and sets steps for progress. Plans would include a process map of steps to integrate BIM into the estate management and development process, the information required to support BIM, the technological infrastructure needed and the education and training required by staff. Policy on selecting suitable suppliers and on use of contract tools would be on the list, as would development of a model BIM Project Execution Plan, the template for procedure.

The guidance to use includes PAS 1192-2 and its forbear, BS 1192.2007. These are the key guides to information management in construction. PAS 1192-3 covers asset information management, the start and end point for Employer's Information Requirements in BIM practice. PAS 1192-4 covers COBie, the tool for converting model data into spread sheets and CAFM tools. PAS 1192-5 advises on how security can best be maintained in creating and sharing data. The reading list includes the CIC Protocol for BIM appointments and contracts, which includes guides to the role of the Information Manager and to the effect on insurances from using BIM. The important fact about Level 2 BIM is that the insurance industry agrees that nothing serious changes in the roles and liabilities of the parties at this level of use.

The top level source of guidance is the BIM Task Group website: www.bimtaskgroup.org. Most of the material is written for supply-side use but advisers will help filter it for client information.

It is not necessary for clients to invest in the level of technology, software or training which designers need to have. Clients can view the model using a normal PC/Mac and software like Solibri or Navisworks that opens models to see and check but not to affect them.

5. How do we appoint teams that can deliver BIM for us?

Consultants and main contractors are at varied levels of maturity in their use of BIM. Few specialist contractors are experienced yet and few product suppliers have put their information into BIM formats. These will all change over the next three years but now it is wise not to select firms without paying attention to their ability to 'play BIM well with others'. It is a collaborative game but one with still-emerging standards for collaboration. One choice is to seek an integrated team with a record of collaboration and BIM usage. This could be a full contractor-led team from the start, with the contractor's choice of designers and specialists. Or it could be an integrated design team put together by the client's chosen architect, with the contractor's team added no later than Stage 3. A 'construction manager' approach would put together a team based on proven compatibility but under client choice and leadership. Other team members, project managers, cost consultants and advisers need to be from firms that can or have worked together as people and

with the tools intended. Central government has primed the market by forming frameworks of suitably qualified integrated teams from which they choose per project.

Clients should put out their Request for Proposals with a clear intention to use BIM at the level chosen. Employer's Information Requirements should be part of the request and bidders should be asked to set out their capability to meet the EIRs. You should request a draft BIM Execution Plan to deliver the required information as a pre-contract basis for discussion. That document should include assessments of team member capability on a template provided by the Construction Project Information Committee (see cpic.org.uk).

Clients have a new job in appointing the Information Manager, usually from within the selected architect or contractor firm and with the role potentially moving from consultant to contractor at Stage 4. There are also BIM consultants available who will play this role from outside the design-construct team.

6. How do we value the costs and benefits of using BIM?

This is uncharted territory as yet. The expectation is that BIM will eventually bring a step-change in productivity in the industry. If coupled with good client and FM skills, risk will be reduced, time saved and costs reduced. As BIM practice advances further, these savings should increase and when coupled with Soft Landings should raise performance and reduce operating cost. It should also greatly reduce the likelihood of defects caused by information error. Facility management costs over the life cycle should come down substantially, as US experience suggests.

As a basis for valuing benefit, the following points are worth considering:

▶ Improved project outcomes flowing from better brief-making, earlier and better decision-making, use of library standards, Soft Landings and feedback from in-use evaluation.

▶ Decreased operating cost flowing from the Soft Landings process, better system performance and better comfort.

▶ Decreased FM costs flowing from easier access to O&M information.

▶ Reduced risk to project quality, time and cost targets from better information management.

▶ Reduced capital expenditure from shorter project time, reduced contingency and greater efficiency.

▶ Earlier availability of the asset from a shorter project timescale.

▶ Decreased risk of claims, defects and legal costs from better process.

As a basis for considering costs associated with BIM use, consider the following:

▶ Costs of setting up client systems and of training staff to work with BIM and COBie, spread over following projects.

▶ Additional cost of advisers at Stages 0 and 1 to define and prepare the project properly.

▶ Front-loaded consultancy costs to make and host models, potentially offset by reduced cost to all parties later in the project.

▶ Additional cost of Soft Landings service from consultants and contractors.

▶ Additional costs of in-use evaluation to gather feedback.

▶ Possible (once proven) cost of integrated project insurance held by the client, potentially recovered from consultants and contractors not needing their own cover.

Overall, costs are likely to reduce more in the build and operate phases than in the design phase. This pattern would parallel the trend in manufacturing where research and design are now a larger part of overall costs, with fabrication and operating costs a lesser part. Higher performance, quality and value can be achieved by this trend, increasing competitiveness for the client and for suppliers. Given that there is expected to be continuing upward pressure on capital costs from world demand for materials, rising sustainability goals, increasing energy cost and skilled labour shortage, the countervailing savings from the use of BIM and its associated processes is a promising way to contain these inflationary forces.

7. Can BIM help us with sustainability?

The sustainability of our civilisation is a major challenge to the current economic model. Market forces are not generally showing the potential to solve this challenge, leading to the declaration by Lord Stern that this is the greatest market failure in history. For clients with a commitment to more sustainable development, the cost of reaching higher standards is a constraint. BIM can help in two ways: improving the design and commissioning process to achieve higher performance more easily, and reducing whole-life cost to make investment in higher standards more affordable.

BIM as a design tool provides the basis for simulating performance. Analytical applications can be run on the model. These can simulate space use, environmental performance, fire behaviour, structural stresses and, doubtless, an increasing number of useful properties. All these simulations can reduce overdesign and increase safety.

The economic advantages of BIM-based working can counter inflation. Clients can choose to reinvest some of these savings into higher building sustainability without sacrificing other goals. Significant whole-life cost reduction is part of the Construction Industry Strategy for 2025, as is capital cost reduction and carbon reduction. BIM is one of the major tools available to achieve more with much less.

8. Can BIM help us with asset management too?

The arrival of BIM marks a switch for the construction industry from considering buildings as projects to considering them as assets. The circular model of the asset life cycle replaces the linear model of a project. For clients with estate or infrastructure assets, the working method of BIM promises to serve them better.

9. What comes next after Level 2 BIM?

Level 2 BIM is required on work for central government departments from January 2016. It will gradually spread to become the basic working tool of the industry, starting as it has already with interested clients in local government, universities and commercial development. Level 2 is an entry-level BIM concept, one which clients and suppliers are expected to use without remodelling any other aspect of their way of working than the information management changes outlined in this guide. There is, however, an expectation that BIM will move onto a more ambitious plane called Level 3 in the 2020s and that many other relevant changes will occur.

Level 3 BIM is expected to centre on shared use of a single model held in the common data environment, rather than the federation of separate models. The CDE will also probably be provided in the cloud rather than on a dedicated server, unless security requirements are overriding. This would enable rapid integration of contributions from all users and further accelerate project progress. Collaborative contracts, which support integrated teamwork but provide less potential to pinpoint liability, are seen as important to effective BIM working. These contracts, notably NEC3, PPC 2000 and JCT-CE, are being used now with Level 2 BIM. The Ministry of Justice is using PPC 2000, a partnering contract. The necessary commercial pressure is provided by the clients' power of patronage to give future work. The whole team stands or falls by its performance rather than being adversarial internally.

Significantly, BIM diminishes the potential for adversarial relationships between clients, consultants and contractors through the following:

▷ Prompt payment guarantees, including the project bank account and removal of retentions. These methods increase the comfort of tier 2 suppliers and levels below, increasing their commitment and protecting their viability. Their prices should reflect that. Lead contractors may seek additional preliminary payments to replace the cash they used to hold back.

▷ IPI: this approach is being trialled now. It replaces the separate insurances of the designers and constructors, which push them apart when problems arise, with a single policy held by the client.

▷ Integrated regulations checking. It is possible to use BIM to manage statutory regulation of design, either by building the approved design standards into design software or by having designs checked automatically against regulations.

▷ Completion of the full menu of interoperability rules: the many types of software used across the industry can talk to each other if they have been written on the basis of a convention called IFC (Industrial Foundation Classes). IFC conventions are well advanced but also require completion of work on an international data dictionary and a standard for processes in data management.

▷ Product information standards: at present product manufacturers do not have a single format for the provision of digital information about their products. A medley of alternatives exists to allow use with varied BIM platforms and to serve the requirements of particular contractors. This costly anarchy needs to be replaced by a standard format.

▷ Internet of Things (IoT): this term describes the use by devices of the Internet Protocol to communicate. It enables devices to supply data into analytical algorithms to support a wide variety of applications. Low-cost sensors are now arriving which can make a building or infrastructure asset capable of being managed as are jet engines and Formula 1 cars: automatic, real-time control to meet changing circumstances, plus full reporting. IoT is likely to join with BIM and Soft Landings as complimentary concepts. The Smart City concept parallels IoT for buildings, offering multi-facetted management of urban areas and infrastructure use.

It is probable that the combination of these potentials will define the next level of BIM, and BIM may come to be seen as a component of wider synthesis of digital asset and city management. Most of the potentials described here will mature before 2020.

It will be obvious from the above that BIM is not a fixed concept but one which will continue to develop with emerging technology and changing market offerings. Clients who start down the road to BIM usage should be aware that the road does not end at Level 2. The outline plan for Level 3 was published in early 2015, under the title Digital Built Britain, as a vision for the next decade of development. Level 2 provides a landing on the grand staircase. ■

The transformative nature of digital

How we deal with clients, as well as how we think of space and place, is being utterly transformed by digital technologies.

Starting with clients, we are moving from a world where it was natural for one party to talk at, rather than to the other, to one where constant, deep and mutually rewarding collaboration and communication with – is the order of the day. Co-creation, user feedback and iterative product development is becoming commonplace. As with major brands and advertising, where blasting out marketing messages is being replaced with ongoing customer dialogue via social media, we are entering a world where monologue is no longer viable. And it is in understanding the effect of these changes that real value can be created. Building a great business today means creating a great user experience. For everyone you work with. Because in this digital world your UX (user experience) is your brand.

The user experience each and every customer, supplier or colleague enjoys in each and every interaction with the client's company is their brand. It embodies everything about them, their product, their values, their culture, their worth, their reliability and their fundamental ethos. Put it all together and it is, as the American investor and business magnate Jeff Bezos says, what people say about you when you're not in the room.

And in a digital world, where bits are more important than atoms, and where people have more choice in where they live or work, and with whom they work, or shop, or play, the client's brand – that essence of them – matters more than ever before. We all want to engage with people we respect, and people we respect in architecture engage with us through well designed and accessible environments.

The great brands of the future will be companies beautifully designed from top to toe. By breaking down corporate silos, tearing technology out of the hands of the IT department and by feeding off the combined intelligence throughout our companies we will build better businesses, products and experiences for our customers.

There must be a coherence of approach, so that everyone understands how the client's organisation works. Dealing with the client's company should be easy for your customer. And the look and feel of interactions should be consistent. Drive one BMW and you can drive them all. That is how working in or with the client's company should be.

Now to space and place, as the digital world is also having a profound effect on how we think of that. Traditionally, we have thought of the space we inhabit purely in physical terms, and all questions have been answered with build this, build that, add this, take away that. But today every place has a digital layer. Think of it like this:

▶ I am here.

▶ What's going on?

▶ What happened here?

▶ What's going to happen here?

▶ Are my friends here?

▶ What did they like?

▶ What should I do now?

People want information; especially local, in the palm of their hand, and physical space has a past, a present and a future. The digital layer is about unlocking this content.

The ubiquity of smartphones, and increasingly fast broadband, opens up this world. Our experience of the space around us is now as heavily influenced by that computer in our pocket as the physical environment in which we find ourselves.

It is the smartphone that will determine how 'a sense of place' is experienced. Think about the questions above; they cover the variables that determine how successful a place is. At the human level, they are what all of us want to know. The place that helps us answer these questions will be successful. The place that informs, inspires and excites us will prosper.

Socially productive places will embrace this digital layer. And if they do not, then they will not be fit for purpose. We can talk all we like about mixed uses, high quality architecture, good urban design and sustainability but if we do not expose the digital layer, then our place is stuck in the past.

Winston Churchill said, 'We shape our buildings; thereafter they shape us.' Today the same applies, but with an added, digital dimension. Embracing this, as well as being an essential business survival strategy, will allow us to build better relationships with our clients, and better spaces and places for all of us in which to thrive.

Chapter 9 Managing and constructing the project

by Tom Taylor

"Client project personnel make a difference. These are the people who work together to push the project along; collectively and individually they have an accurate view of the project overall and its prognosis for satisfactory delivery. They anticipate problems, and if they cannot avoid them, they try to solve them. They understand blame but that is not their first reaction."

Introduction

This chapter refers to a project as though it is actually on-site. This is the busy, exciting period; or even busier and even more exciting than before – if that is possible. There will be lots of activity, by lots of people, on- and off-site, spending lots of money every week and no going back. What does a client do now? Despite all the busyness, the official answer is probably 'not a lot' except pay the bills and make the relatively minor decisions on issues such as paint colours. So do not make changes, enjoy the experience, be enthusiastic and appreciative, and wait patiently for your completed building and project to be delivered. This is when everything should come together – in a physical form – so some pre-site as well as on-site issues are incorporated in this chapter.

The project on-site – changing responsibilities

As the project moves onto site, so the site responsibilities change over. This can be an unexpected shock for some clients. They still own the real estate but responsibilities move over to the principal contractor – for security, access, insurance, fire precautions, and the safety of workers, visitors and the public – as well as for executing the works.

Clients, their staff, advisers and their own direct contractors need to understand and respect these temporary arrangements. They may need specific briefings and instructions –

by the contractor. It may also be appropriate to have some joint walk arounds, briefings and handovers with the client – during the mobilisation period between placing the primary order as the contractor takes possession. If the client does not feel sufficiently confident or knowledgeable about their own buildings or site, they can ask the confident and knowledgeable design team and property advisers to undertake the role. If a design and build approach or managed form of contract has been adopted, then a pre-construction period with greater involvement of the contractor may have occurred with greater opportunities for buildings and site familiarity.

Going to site

Having relinquished responsibilities for the site for the duration of the works, it is important that the client understands the protocols of going to and visiting the site.

Here are some tips for clients:

▶ Make an appointment or receive an invitation – do not go unannounced, it is discourteous.

▶ Go for a purpose, identify the purpose, achieve the purpose. Have more than one purpose, have a list, do what is on the list, check what is on the list.

▶ Be prepared to be inducted on your first attendance on-site. Allow time, pay attention and set a good example even if you have been on construction sites before.

▶ Sometimes it is possible to go to the site without going on-site, for example to attend a meeting in site accommodation or nearby without the necessity of setting foot on site; do not go for one near-site purpose and then adopt another on-site activity.

▶ Become familiar and comfortable with personal protection equipment (PPE) of footwear, jackets, helmets, protective eyewear and gloves to suit the situation. If site visits are to be a regular activity, think of getting your own PPE with your own logo (and not someone else's), which you keep or leave stored securely on-site.

▶ Always tell people where you are going and when you should be back.

▶ Always sign in and sign out.

▶ Always take advice and directions when on site.

▶ Be observant on-site for your own safety and of others. Most accidents happen through lack of familiarity; and a feature of construction sites is that they are constantly changing.

▶ Always set a good example in all such matters – after all you are the client! Overfamiliarity can lead to complacency and increased danger. This is, of course, an extreme example.

Meetings on-site

Meeting arrangements are different when the project is on-site and construction is underway, in frequency (more often), in regularity (some more regular plus others more spontaneous or individual), in attendees (more participants and from different sources). More people and organisations arriving and being involved seem to mean more meetings!

The client will need to carefully consider which meetings to attend or at which to be represented or not. It is good to resolve such matters from the start – and yet be flexible enough to optimise commitments, contributions, information and presence.

If the client is a stranger to site meetings, it is wise to take it steady – especially initially – as an interested observer or attendee. Sit next to someone who can be a guide or buddy; leave any items or concerns until 'any other business'; be prepared for a certain amount of banter between parties. Decide what to do about jargon – whether to ask for clarification sooner or later. Expect differences – refreshments may be tea and bacon rolls rather than coffee and biscuits or croissants (or possibly vice versa).

Project managers or managers of projects

There are some subtle differences between being involved as a project manager and contributing to the management of the project. Client project personnel make a difference. These are the people who work together to push the project along; collectively and individually they have an accurate view of the project overall and its prognosis for satisfactory delivery. They anticipate problems, and if they cannot avoid them, they try to solve them. They understand blame but that is not their first reaction. They may be seen or heard conversing before or after meetings, in corridors and car parks, talking about the project (not everything is resolved in formal meetings or by emails).

This group exists on most projects; some are much better and effective than others. Clients need to be aware of such groups and decide how to engage with them individually and possibly collectively – and how to participate. Generally these people operate at project level and so they are suitably project knowledgeable – not too senior, confident, not too junior. The good participants are those with whom the client would like to work with again. This is not a 'steering group', it is a collection of 'drivers'. Sometimes it is all subconscious. It is a form of selective, project-focused collaboration.

Value engineering and value sourcing

Value engineering attempts to achieve additional value for the same or little additional cost; or attempts to achieve the same value for less cost, usually achieved through scrutinising functionality, technical specifications and quantities (as units, volumes and areas). The ultimate target might be to achieve greater value and also reduce costs – sometimes summarised as 'even more for even less'.

Value sourcing is different and is all about trying to obtain the same value at a lower cost by seeking alternative sourcing of goods, products and services or by contract arrangements – which may also lead to different technical solutions.

Value engineering might be summarised as 'make it cheaper', while Value Sourcing might be 'buy it cheaper'.

When the team is undertaking value engineering and sourcing in order to achieve budget reductions, expect the need to identify savings equivalent to twice the gap because only

half will usually be deliverable. Therefore go for big dramatic effective possible savings rather than many small possibilities. Also identify, tackle and engage with the 'quiet' participants in such exercises. They, too, can contribute.

Change – good and bad

Capital construction projects are all about change. That is good. Projects take a situation or status quo, manipulate it – usually over a relatively short period – and then deliver an improved status quo for the benefit of the customers, users, society, civilisation, etc. over a long period.

The construction project by itself does not usually provide all the changes required. People, equipment, technology and new ways of working will also be necessary. A physical hospital building contributes to a health facility, a school building is part of an education establishment, a house becomes a home, and so on.

And yet there is great sensitivity and concern about 'change' taking place within and during projects. Within the individual stages of projects good ideas, development and refinement are encouraged to suit the stage of work. This is sometimes called iteration. The difficulties occur when changes are identified or requested that are out of sequence. They are being requested at an inappropriate stage – usually too late – with the consequence that such potential changes can be disruptive to the timetable, the budgets and the holistic solution if adopted, and even if considered but not adopted.

Here are some observations about dealing with change on projects:

▶ Have a change management system or protocol for processing the assessment and decision-making of potential changes.

▶ Also have a change resistance mentality, which stops topics getting into the change management system.

▶ Differentiate between 'nice to have' and 'must have', and then apply common sense.

▶ Changes can result in savings (sometimes) as well as extras (more often).

▶ Establish if the change alters the signed-off brief, the signed-off designs, the signed contract, the works themselves or some other aspect. It is surprising how many changes have their origins earlier in the previous project processes – going right back to within the initial brief or feasibility studies or budget or scope assumptions.

▶ Make financial provision for possible changes (a good idea, usually as a contingency sums) within the contract sum and/or within the overall project budget.

▶ Categorise all changes, perhaps on the basis of client changes, design development, procurement influences, regulation and statutory requirements, site conditions and other combinations.

▶ It is possible to include some simple credits that can be used to pay for extras, for example with higher specifications or functional provisions or provisional sums, or day work provisions.

▶ When a change is identified there can be agreed review routes on the lines of:

• A full comprehensive report with implications, alternatives and recommendations, about ten working days to produce, costing say £5,000 to £10,000 to produce if paid for or not, and disruptive to on-going activities in its production even if not implemented.

• A simple short assessment, taking about five working days to produce, at a cost of say £1,000 to £2,000, with less disruption to activities.

• A 15 to 20 minute discussion within the team members present at the time with the originator of the request. This process can often squash a changed idea or considerably refine it to enable one of the other approaches to be adopted effectively. Additionally, this may stimulate requests to be thought over before being expressed – since they are going to be openly and positively cross-examined.

▶ Maintain a schedule of change assessments and monitor the number, status and decision-making – managing accordingly.

▶ Undertake 'lessons learned' exercises at the end of the project against the whole change schedule to assist in improving the accuracy of predictions and efficiency of processing.

▶ Expect some 'extras' to be twice the expected cost and some 'savings' to be half the expected saving initially – interrogate and negotiate accordingly.

Expect people to be 'helpful' in raising a change topic by providing the answer rather than the question, eg they will request, 'We need to widen the door opening. How much will it cost?' without saying why. Fundamentally, the client asks questions and the project team provides options and answers. The same should apply to change items.

Expect some 'new' ideas to not appreciate that they may be covered by existing provisions and/or have particular justifications – sometimes good memories, project commitment and a steady nerve can be of assistance.

Construction programmes and project programmes

Usually when the project is on-site the predominant programme will be a detailed construction programme prepared by the principal contractor covering the contracted works within the contracted start and finish dates. This can be a very useful document usually represented as a bar chart, preferably with a method statement that is a commentary on how the project is to be constructed.

However, usually not everything that needs to be done on the project – and that is of interest to the client – is certain to be reflected in the construction or works programme. These omissions are to be expected and need to be identified, tackled and monitored by the client and their advisers – because they may be their responsibilities – and be vital to the project overall.

These further topics may include:

Off-site works and fabrications with inspections – frequently construction programmes only cover the on-site works.

Placement of orders to and works by utility suppliers.

Orders and installation of client-supply furniture, fittings and equipment (FF&E), which may overlap with or immediately follow the construction works.

Celebratory events such as initial start-up, turf cutting, weather tightness, topping out, first phase completion, overall handover, occupation and official opening.

Release of provisional sums, clearance of remaining design topics, resolution of samples/mock-ups/first rooms.

The selection of off-site sourcing of raw materials, processed materials components, sub-assemblies and whole units will affect the on-site timescales and costs of capital works but also impact on the running costs thereafter.

Observing the works in progress

To begin with, work can seem to be going backwards. More material may seem to leave site than arrive on-site – such as demolition arisings and excavated material. Existing buildings can be demolished or stripped out. With basic foundations and certainly with new basements, the project may be seen to be going downwards rather than upwards. So in due course it can be a key point on the programme when the project 'comes out of the ground'.

Construction sites are busy places, usually with multiple activities taking place at the same time. It can be difficult for clients to monitor the progress and assess if the project is going to be on time for completion.

The specialists in such matters can provide reports with assessments of progress against programme or timeline. Different approaches can be adopted including:

▶ Critical path monitoring, remembering that there can be many near-critical activities.

▶ Activity counts – as in how many activities completed per week/month against as planned.

▶ Milestone achievements.

▶ Activity starts – for trades. Trade starts are easier to identify than trade completions.

▶ Financial valuations and payments against predicted cash flow.

Clients should consider standing still in the same places on-site every time they visit and look at the view. This way they will be able to see progress. Take a photograph or two – even a selfie with the project in the background. Your clothing and expression will indicate the time of year, the weather conditions and your opinion on progress.

Quality

What is quality? Is 'quality' the best word to use?

Is it better to consider 'performance' of which quality is a part, along with scope, appearance, aesthetics, consistency, reliability, technical achievement, statutory compliance, tolerances, maintainability, robustness, removability and others? When there is a problem, disagreement, dispute or misunderstanding on quality, it is usually about one or more of these criteria. Which ones? Let's get the right ones on the table.

Pre-agreed standards and benchmarks sound like a good idea but they need time and effort to explore and agree. And they need to be pre-agreed – not as the problem arises and thus become part of the problem.

Table 9.1 (opposite) provides a menu of possible means of explaining the quality and performance expected or desired. In practice, the menu can be applied to establish who will produce the quality criteria or material and who will receive them, use them or rely on them. For example, the specification included in the contract documents is often thought to be a key means of conveying standards and requirements for workmanship and materials, however few see tradesmen consulting copies of the specification on-site. What are they using?

How to use the Quality Menu:

▶ Apply to specific projects Add title, date and the current stage.

▶ Identify key parties for quality across the columns eg client, designers, principle contractors, suppliers, tradesman.

▶ Identify generic roles within the columns, eg originator (O), checker (C), user (U), tester (T), inspector (I).

▶ Work up the detail of topics, tasks and who does what.

Quality Menu: [Project] [Date]					
Quality participants	C	D	PC	S	T
In the brief					
By common understanding from visits to relevant examples, benchmarking					
In the selection of team members					
By establishing quality as part of value within budget provisions, overall and within elements					
In the outline proposals (C) and detailed design/scheme design reports (D) or Concept/Developed/Technical Design (Stages 2, 3, 4)					
In models, artwork, illustrations, computer-generated images (CGI)					
In shortlisting selection of contractors and suppliers – and their responses					
Within tender documents					
On for-construction drawings and specifications – within fulsome Building Information Models					
By individual material samples					
By referencing and consulting British Standards, Codes of Practice, other standards					
By assemblies and prototypes					
Through consistency across the project and its uses for its life cycle					
Through the selection of trade workforce					
By suitable tolerances					
During sourcing, manufacturing, assembly, delivery processes					
With on-the-job Quality Assurance procedures					
By special inspections and testing, eg for insurance, welding, concrete sampling, BREEAM criteria					
By programme sequences, durations, methodologies – on- and off- site					
By protection of work in delivery, in progress and completed installations					
By progressive and final inspections, testing, commissioning and snagging					
With handover briefings and documentation (selected BIM)					
With post-completion support when in use					

☐ Table 9.1: Quality Menu blank

KEY:
C = Client   PC = Principal Contractor
D = Designer   S = Suppliers
CGI = Computer   T = Trade (subcontractors)
Generated Imagery

Quality Menu: London Academy 3 March 2016					
Quality participants	C	D	PC	S	T
In the brief	0	O/I	U/C/T	U	U
By common understanding from visits to relevant examples, benchmarking	0	0	0	0	
In the selection of team members	0	0	U	U	U
By establishing quality as part of value within budget provisions, overall and within elements	0	U		U	U

☐ Table 9.2: Quality Menu example

Table 9.2, above, provides a short example of how the table may be used.

When clients get into conversations about quality and performance – especially when the project is on-site – it is important that there are clearly established criteria and these are to hand, rather than a memory of the supposed criteria or assumptions or opinion alone. This attached listing or Quality Menu can assist.

Testing and commissioning

Frequently the primary representation for testing and commissioning is a period at the end of the contractor's construction programme between the end of most of the works and the designated completion date.

This is an important period. It is not float in the timeline. It should not be allowed to be squeezed by late running works while still attempting to maintain the dates for completion/handover/occupation. An under-tested and under-commissioned project will almost certainly be regretted later.

Furthermore, the testing of materials, components and subsystems will usually have been taking place throughout the works period together with some commissioning of installations off- and on-site. Consequently, the testing, commissioning, plus likely inspecting and snagging are not just concentrated into the final testing and commissioning period. Clients need to agree with their teams when they or their representatives will need to be involved with or participate in testing and commissioning throughout the project period on-site. Soft Landings are also long landings.

What is a project execution plan?

A project execution plan (PEP) may be called by a number of other titles, which are all subtly different and in practice mean what the users choose them to mean, for example the project plan or the project management plan (PMP). Client's requirements should be expressed in a brief, statement, wish list, Project Initiation Document (PID) or Business Case – in any document that explains and justifies what is required: the definition. Some say this is the project execution plan.

A good PEP will be a clever combination of the strategic, the specific and the tactical. It is as well to discover what the team means when it uses such terms. If there is a PEP, it is advisable to look at it regularly, make sure it is used and keep it up to date.

In conclusion

Every construction project in its circumstances is unique. Every development and resulting facility is unique. Every team of participants and stakeholders during the construction phase is unique in its composition and relationships. Most of these participants and stakeholders want the client's dreams to come true rather than become nightmares. ∎

Characteristics of a good client

The need for architectural and associated design services frequently coincide with periods of fundamental change within the client environment, for example a major relocation or head office redevelopment. Such change can create a period of business uncertainty and staff instability. However, these kinds of risks of possible disruption are usually justified by the prospects of long-term benefits such as improved productivity and corporate positioning. One such example was the amalgamation of 5,000 Reuters London staff from seven disparate buildings into a new purpose-built office in Canary Wharf in 2006. This created a more productive atmosphere with closer linkages, up-to-the-minute technology based tools, plus a 30% reduction in office costs.

The overwhelming message from some 30 plus years of major change programmes is that beneficial outcomes can only be achieved if professional service providers – ie architects – and the clients are prepared to share the risks and challenges of a turbulent journey, as well as the intended benefits. In so doing, the soft factors that can mediate against successful programmes may be identified and dealt with by both parties in an equitable way. Clients need to accept that it is not possible to outsource all risks to a third party merely because the contract 'says so'.

BP is a company that appears to encapsulate best practice here. One classic example of how a provider and client can work together to achieve remarkable outcomes occurred some years ago when BP Europe chose to bring all its office services together into a shared support centre. The project involved the amalgamation and centralisation of support services in 26 different countries – each of which had independent processes and skills employed locally to provide such services. The business goal was to halve support costs whilst improving staff perception of service quality (from an average of 3.4 to over 4.0 out of a total score of 5.0).

Initial efforts by the successful service operator and change partner (Fujitsu Services) met with growing hostility by many of the national BP affiliates as support services were extracted from the local environment and relocated to a shared centre in Holland. At the low point, the BP executive in charge informed Fujitsu that he was prepared to cancel the entire contract. However, sanity prevailed and the two companies sat down together to discuss how they could achieve a mutually beneficial outcome. Each accepted that blame needed to be apportioned to both sides. Neither should be prepared to resort to contractual wrangles. BP established a special task force to visit all

26 countries and begin a programme of staff education and orientation, geared towards promoting the benefits of a shared environment. Fujitsu relocated BP's centre from Holland to Portugal to create an entirely new operation that had been purpose-designed for this demanding and multlingual European BP community. Within three years costs had been halved and customer satisfaction increased by 20% from the initial baseline. Everyone agreed that the programme had met its target, and the contract with Fujitsu was extended for several further years.

One of the most likely points of failure in such large scale and complex programmes is the contract itself. Many successful programmes have required both client and service provider to tear up the original contract and work on a more informal basis to achieve a successful collaboration.

This has been particularly the case in the public sector. For example, one of the UK's largest government departments came within a hair's breadth of launching a £10bn lawsuit against its primary IT vendor when a ten-year contract came unstuck. Poor initial contracting combined with overweight public scrutiny of ongoing services brought the two sides to a hostile stand-off. The situation proved so unsatisfactory that the entire IT sector was called to a meeting with the department heads. Any lawsuit would have endangered not just the prime contractor but a host of other engaged parties to the extent that the entire industry could have been crippled for years to come. The interesting outcome of this meeting was the creation of a joint task force to review not just IT but every aspect of the department, from property and energy consumption to staff productivity and automation. A report was prepared by five leading IT suppliers and submitted to the board. This invoked a full renegotiation of the existing ten-year, £10bn contract, as well as a strategic review of the business itself. As a consequence, the IT spend was reduced by some £350m per annum and the staffing of the department was cut from over 140,000 to less than 90,000 in just three years. One can only conclude that a dispassionate and fully collaborative approach will always yield far greater gains than a legal conflict.

In summary, the history of large change programmes and outsourcing contracts suggests that legal contracts alone can rarely – if ever – accommodate or stand up to the many changes in circumstance that large, complex organisations undergo. Instead, all parties need to find a mechanism for engaging in a 'no-blame' collaborative atmosphere where collective interests can be rewarded on a continuous basis, and an atmosphere of risk-reward is the primary driver. Many companies today describe this as a process of innovation and creativity that lends itself well to the new social and economic environment of the digital age.

Chapter 10 Accepting, commissioning and using the project

by Gary Wingrove

"Planning the move of the business and its people is just as critical as the build project itself and can take as much thought, time and preparation."

Introduction

The transition from the build phase through commissioning and acceptance and then into the day-to-day use of the building is critical and if not planned for and carried out efficiently, can undo all the good work that has gone before. The subsequent inefficient running of the building will be a waste of the time, knowledge and money invested, and will result in higher in-life running costs. So the preparation to take ownership of your project should start with the defining and designing of the project and carry on through the build phase.

'Practical completion' is the phrase most widely used when describing the point at which the construction is finished and the occupation begins; this phrase is a contractual term used in many (but not all) forms of construction contract, and because of that the term 'handover' will be used in this chapter.

Leading up to handover

The most effective and successful way of achieving a smooth handover is to plan this from day one. If at all possible, those who will be running and managing the buildings should have been involved in its design to ensure that they fully understand its working and get the most out of it. Some of the key areas that should be discussed and agreed at the design stage are:

▶ Plant replacement strategies.

▶ Choice of easily maintainable or replaceable finishes.

▶ Choice of equipment and fittings that are easily replaceable and fully warrantied.

▶ Agreement to the type and complexity of the building management system (BMS).

There will be a hive of activity on-site leading up to the handover; the contractor will be trying to complete the works, the consultants will be inspecting the works for completeness and quality, and will be producing snagging lists, which are schedules of work that in the eyes of the contractor are complete, but in the eyes of the consultants are not, or are not to the right quality. The consultants will also be heavily engaged in the testing and commissioning of all the systems to ensure that they are performing as they should and, in particular, that all resilient and emergency systems react as they should in the event of, say, a major power failure.

It should always be an aspiration to achieve zero defects at handover, and indeed a number of contractors now adopt zero defects processes so it is certainly worth favouring a contractor who operates with these schemes. It shows they are committed to the handover process and often invest in people and technology to achieve their goal, which is a definite benefit to you, the occupier. It can be very disruptive having contractors still on-site after occupation and can detract from what otherwise should be a positive experience for all involved. It is also worth remembering that under most forms of contract there will be a defects liability period. This is typically 12 months and is effectively a guarantee that should anything go wrong with the project in that period, other than fair wear and tear, the contactor will put it right.

What happens at handover?

The day of the handover is when the client and the consultants agree that the project is practically complete or, in other words, is sufficiently complete for the client to take occupation. The ultimate decision is that of the clients but in practice this is often left up to the client's project manager and design team as they will have the necessary experience and should be protecting their client's interest. The day of handover is one of the most important days in the project lifecycle and is also a contractual requirement which has positive financial implications for the contractor, but do not be pressured to accept a project you are not happy with, or compromise on the quality of the finished product.

The handover day must be properly planned to ensure that it runs smoothly and that all parties are aware of the agenda and their responsibilities. Any outstanding works should be agreed and scheduled prior to the handover day. The contractor and consultants should have produced a handover checklist and this is the starting point for the completion plan. This should form the agenda for the handover day, and as a minimum it should include:

- ▶ Design and documentation reviews.
- ▶ Agreement to the construction programme status.
- ▶ Inspection and snagging, and agreement of outstanding works programme.
- ▶ Testing and commissioning.
- ▶ Statutory inspections.

- ▶ O&M manual and handover manual.
- ▶ Health and safety file.
- ▶ Spares.
- ▶ Warranties.
- ▶ Client training.
- ▶ Environmental assessments and documentation.

What are the client's responsibilities at handover?

Beyond the responsibilities of the client on the day of handover there are several other responsibilities that must be addressed as the client takes occupation of the building or space. If leasehold arrangements exist, then the occupation of the building may, if the client is not already doing so, activate the payment of rent or the rent-free period. Clients will also need to make sure that the necessary insurances are in place, whether provided by them or by the landlord. Business rates will also kick in on occupation, and arrangements to pay these need to be in place.

Finally, the client is responsible for the health, safety and security of the occupants of, and visitors to, the building or space so arrangements need to be in place to manage security, either physical – such as reception or manned guarding (or passive) which would include alarms and CCTV. In addition, first aiders and fire wardens need to be appointed and emergency procedures agreed and communicated.

Alternatives at handover

By far the most common situation is to take occupation of a building on a given day. However, there are situations where it is beneficial to the client to agree additional occupation dates, or agree on a phased approach, the most common of which are listed below.

Early access

This is used when the client requires access to the building before formal handover in order to carry out their own works. Typically, this will include the building of Comms Rooms, specialist fit-outs such as audio-visual suites, and catering facilities. These works are normally carried out under a Permit to Work process as the contractor still has responsibility for insurance, coordination and health and safety on-site.

Phased completion

Adopting a phased completion approach, typically where one or more areas are finished and handed over to the client in advance of the main works, can often reduce the project duration. This 'partial possession' approach may be attractive when business constraints, such as overcrowding in their current space, required an early move to mitigate the need for a short term let, or where swing space is needed to facilitate the move management. It is by no means a perfect solution, as insurance, health and safety considerations, security of the completed areas, access, dust and noise all have to be considered, but the advantages may well outweigh these concerns.

Soft Landings

Soft Landings is an approach that contracts the designers and contractors to stay involved with the project beyond handover and to work with the client and his building management team to fine-tune and ensure the efficient running of the buildings systems.

Moving in

Planning the move of the business and its people is just as critical as the build project itself and can take as much thought, time and preparation. If the move is not part of the main project, then it is advisable to engage a separate move management company who will plan and manage some, or all of the following.

Housekeeping

Only move what you need to, especially if you are changing working practices – for instance moving to a paperless office. Filing audits, etc are recommended and it is a time to unclutter.

Communication

This is probably the single most important element of the move process. People will need to be guided through the whole process, so continual communication stating what is going to happen when is essential.

Coordination

It is critical that no services fall between the gap of the main project work and the move management process. It needs to be clear who is responsible for moving or transferring:

- Catering.
- IT.
- Updating/renewing company intranet.
- Phones.
- Security.
- Post room.
- Reprographics.

Staff inductions

It is important that your people know how to use their building efficiently and effectively, and it is worthwhile investing time in staff inductions – everything from building evacuations to security, catering, waste/recycling, wayfinding, and reporting building faults, etc. It is often useful to appoint floor or team champions to help with this, and inclusion of key information on the company's intranet is essential.

Getting the best from your new building

The choice of building management company or FM provider is critical to the ongoing successful occupation of the building and should be engaged as early as possible in the project lifecycle and included in the design if possible. Their involvement in the handover process is crucial, and the setting and monitoring of stretching Key Performance Indicators (KPI) is critical.

Post-Occupancy Evaluations (POE) are now commonplace and are an excellent way of measuring how effectively the project design is functioning and how well the building is being utilised and meeting the business's need.

Conclusion

To summarise the recommendations of this chapter:

- Start planning for the project handover on day one of the project.
- Set the agenda for the handover day well in advance, and work towards it.
- Get the building management, or FM teams, involved as soon as possible.
- Do not get pressurised into accepting a building that is not ready.
- Do not forget all the other elements and services that do not sit within the main project works.
- Get the people bit right, as a successful people move will turn a good project into a great one; get it wrong and you will be on the back foot. ■

Department of Health – my experiences with the built environment

Working with the public sector presents a myriad of challenges that are not present in the private sector, from the need to hold transparent procurement processes to civil servants who are afraid to break away from a comfortable and measurable past. In my opinion, the first of these had the unintended effect of dampening innovation. Bidding firms were reticent to put into a public domain content that was proprietary. Also, the default position of the public sector is to stray away from new, different or untested ideas. There is inherent risk attached to implementing new ideas.

In healthcare, firms were then faced many times with an intransigent civil service, an unengaged medical staff, and through central dictates from the various departmental fiefdoms, a intransigent expectation of what the outcome should be, usually based on 1960s healthcare provision. Because of these internal structural blockages, it was difficult to reach through to the new and innovative healthcare designs that were being rolled out by international architectural firms. This translated into smaller UK firms winning many of the initial PFI contracts, as government officials avoided what they saw as potentially career-limiting decisions. Also, many health economies had moved to a place where they were building facilities that addressed future healthcare advances as much as they were the present needs. For example, many US architects were designing operating theatres, which could in the future accommodate robotic surgery equipment.

As PFI matured, and government took more of an output-based view of the process, new projects began to allow the introduction of newer and more efficient designs. I think this was also a result of the government beginning to treat the patient as a consumer. NHS providers could see that in an era where taxpayers were demanding better services, based on information of performance easily obtained on the internet. The government was forced to accept that old practices and the status quo were no longer acceptable.

The new reality of patient consumers is also beginning to change the behaviour of doctors. Younger physicians are far more likely to embrace change in all aspects of clinical practice and welcome innovative healthcare building based on new treatments and care pathways.

PFI has left a rich and very visible history of how government has engaged designers and architects over the past decade and a half. Many early PFI projects reflected the buildings they replaced because this was the safe option for government. Recently, however, many of the designs that have become reality reflect a new era in healthcare that addresses patients' needs today and respects the pace of change in healthcare and needs of the future.

It was always clear that the firms who supported the NHS built environment wanted to introduce the new designs far earlier than the client allowed them to. It is also clear that through perseverance, those necessary changes are being reflected in today's healthcare facilities.

I am in little doubt that innovative and involved architects are at the front line of medical innovation, and introduce change through those innovative designs by breaking down walls both built and figurative, replacing them with healthcare facilities that support today's and tomorrow's patient.

Chapter 11 The client and the law

by Murray Armes

"…the (CDM) Regulations recognise that the Client has influence over how the work is done and as such the Client cannot simply take a passive role"

Introduction

This chapter provides a general summary of the law relating to clients and an equally general guide as to what a client should do if disputes arise on projects. Laws of one sort or another govern all construction projects. Effective clients are obliged to understand that they are subject to statutory laws, the breach of some of which can lead to criminal sanctions. The relationships with the various members of the construction team will be governed by contracts and subject to the civil law regime.

Part 1 – Statutory Duties

Clients are compelled by law to comply with various Acts and their requirements including those described below.

Planning, conservation area and listed buildings consents

Almost all construction projects will require planning permission. Certain types of minor alterations can be made without the need for planning permission, and this is known as permitted development.

The Town and Country Planning Act 1990 codified the planning regulations applicable to England and Wales. Part III of the Act 'Control Over Development' effectively gives control over significant construction or demolition operations to the democratically elected members of local authorities. Even if the client employs a team of professionals it remains the client's responsibility to comply with all the relevant planning rules including making the all-important planning application. A condition of consent may be the requirement to enter into a Section 106 Agreement, which may take the form of a financial contribution made to facilitate local facilities and infrastructure. When applying for planning permission, if the development is located in a conservation area, separate conservation area consent will also need to be applied for.

Buildings of particular architectural or historic interest may be listed. This applies particularly to older buildings but also to architecturally significant newer ones as well. There are three grades of listing:

▶ Grade II is the most common[1] and is for buildings of national importance and interest.

▶ Grade II* is applied to particularly important buildings.[2]

▶ Grade I[3] is for buildings of exceptional interest.

Listed buildings can be altered, extended and changed and even sometimes demolished but a separate consent is required. Unauthorised work to a listed building is a criminal offence.

Although not part of planning legislation, any development will have to comply with the Rights of Light Act 1959 and the Party Wall etc Act 1996. Any development will also be required to comply with environmental regulations and laws. There may also be local policies applicable to an area or particular site which regulate development in order to protect the environment and wildlife. A detailed explanation of all these is beyond the scope of this chapter.

Building regulations

All buildings in the United Kingdom must comply with Building Regulations. The Building Act 1984 is the primary legislation under which Building Regulations are made, the latest of which date from 2010 and continues. The Regulations are supplemented by guidance in the form of 14 technical Approved Documents. Compliance with the Approved Documents is not mandatory and those carrying out building works are free to depart from the recommendations but will be asked to demonstrate that their approach at least meets the requirements of the Approved Documents, which if not met will result in an enforcement notice and possibly criminal sanctions. Although most clients will employ construction professionals to design their buildings and to make a Building Regulations Application on their behalf, it is ultimately the responsibility of the client to ensure they do so.

Health and safety

Constructing buildings can be hazardous and 1994 the Construction (Design and Management) Regulations 1994 introduced legislation to deal with safety in the construction industry. Since then the regulations have evolved and the latest iteration came into force on 6 April 2015 which includes more onerous requirements in respect of clients, including the change from qualified to absolute requirements. Failure to comply with the Regulations is a criminal offence and subject to criminal sanctions.

The main changes affecting clients include the replacement of the CDM coordinator with the role of principal designer, which was a reaction to clients who appointed the CDM coordinator too late for it to have any real effect. By making a member of the design team responsible, both the designer's duties and obligations have changed considerably. It is too early yet to see how designers will react to this additional workload and how they will charge the client for doing it. One way may be for the consultant to appoint a principal designer as a sub consultant. For clients who appoint one team to design and one team to construct a project a question arises as to whether or not the role of principal designer can be transferred part way through a project.

Due to the hardening of the clients obligations it is worth reviewing here the text relevant to the changes between the 2007 and 2015 Regulations:

Regulation 9 of CDM 2007:

*(1) Every client shall take **reasonable steps** to ensure that the arrangements made for managing the project (including the allocation of sufficient time and other resources) by persons with a duty under these Regulations (including the client himself) are suitable to ensure that :*

(a) Construction work can be carried out so far as is reasonably practicable without risk to the health and safety of any person;

(b) Requirements of Schedule 2 are complied with in respect of any person carrying out the construction work.

*(2) The client shall take **reasonable steps** to ensure that the arrangements referred to in*

paragraph (1) are maintained and reviewed throughout the project.

Regulation 4 of CDM 2015:

*(1) A client **must make** suitable arrangements for managing a project, including the allocation of sufficient time and other resources.*

*(2) Arrangements are suitable if **they ensure** that :*

> *(a) the construction work can be carried out, so far as is reasonably practicable, without risks to the health or safety of any person affected by the project;*

> *(b) the facilities required by Schedule 2 are provided in respect of any person carrying out construction work.*

*(3) A client **must ensure** that these arrangement are maintained and reviewed throughout the project.*

The new Regulations are intended to make clients much more accountable for the impact of decisions on the health and safety of their projects. Whereas under the 2007 Regulations domestic clients were exempt under the latest regulations they are not and the client, whether domestic or not must appoint a principal contractor and principal designer. Failure to do so means the designer in control of pre-construction works is the principal designer and the contractor in control of the construction phase is the principal contractor. Whilst it may be possible to identify the pre-construction designer it is not necessarily so easy to identify the principal contactor from amongst the array of subcontractors to whom the works may be let.

While the client now has obligations in respect of the Regulations it is not expected that the client will take an active role in managing the work and this is set out in the L Series guidance attached to the Regulations. There is not space in this brief guide to deal with that in detail but the Regulations recognise that the client has influence over how the work is done and as such the client cannot simply take a passive role. For the client, the regulations are about selecting the right team at the right time and assisting them to work together in a way that promotes safety during construction. The client must perform a number of tasks:

▶ Appoint the right team: when appointing consultants and contractors clients should aim to appoint those with a verifiable track record, and those that can demonstrate they have the resources to do the work properly.

▶ Appoint the principal designer early and preferably before any pre-construction design has taken place.

▶ Allow enough time: Rushing a project is a recipe for both poor quality and unsafe working practices.

▶ Provide information: the client is under an obligation to give information about the brief, how the building will be used and to pass on existing information about the site, structures or buildings and any known information about hazards, such as asbestos.

▶ Assist with communications, cooperation and management: the client should ensure its team both communicates and cooperates. During the design stage the team should be encouraged to discuss issues of buildability usability and maintenance. The best way to do this is to ensure that the team attends regular design team meetings, with the client included.

▶ Ensure the site has adequate welfare facilities: the contractor should do this but ultimately it is up to the client to ensure it does.

▶ Ensure workplaces are designed properly: if the project is for a place where people will work, it must comply with the Workplace (Health, Safety and Welfare) Regulations 1992.

It is too early yet to see exactly what impact the new Regulations will have on consultants, contractors and of course clients. However the Good Client should seek advice on their obligations early in the life of a project to ensure they do not fall foul of the draconian penalties that might be applied if the regulations are not properly implemented.

Part 2 – Contractual Duties

To implement projects the effective client will inevitably employ others and enter into contracts with them.

Professional appointments

It is likely that clients will need to employ construction professionals. Different projects require different teams, typically comprising an architect, a structural engineer, a mechanical and electrical services engineer and a quantity surveyor, together with a principal designer.

Larger or more complex projects may need a project manager or employer's representative.

Professionals should be engaged using formal agreements. Most professional bodies have appropriate forms of appointment and require through their Codes of Professional Conduct that these – or something similar – are put in place before any services are provided. To avoid any serious misunderstandings, the client must insist that an agreement is created, carefully reviewed and signed before the professional starts delivering services.

A typical example of what an appointment document should contain is included in the RIBA Code of Professional Conduct, Guidance Note 4.2 that states:

'When contracting to supply architectural services, the terms of appointment should include:

▶ A clear statement of the client's requirements.

▶ A clear definition of the services required.

▶ The obligation to perform the services with due skill and care.

▶ The obligation to keep the client informed of progress.

▶ The roles of other parties who will provide services to the project.

▶ The name of any person(s) with authority to act on behalf of the client.

▶ Procedures for calculation and payment of fees and expenses.

▶ Any limitation of liability and insurance.

- Provisions for protection of copyright and confidential information.
- Provisions for suspension and determination.
- Provisions for dispute resolution.'

If the client uses an unamended standard form of appointment, there are few things he/she should look out for:

- **The net contribution clause.** A net contribution clause limits the professional's liability to that proportion of the loss it would be reasonable for it to bear taking account of its role in the project. If a defect arises in a construction project, there may be more than one party who is liable. For example, poor workmanship may be the result of poor design or breach of contract by the builder; however, under common law the client can decide which party it wants to pursue and in practice it will be the party who is best insured.

- **Limitation of liability.** In common law a professional will be liable for breach of contract for six years from the date the breach occurred, or 12 years if the appointment is written as a deed. In tort the professional is liable for a period of three years from the time the breach was discovered, with a long stop of 15 years. Clients must decide whether the appointment is to be signed as a simple contract or a deed. Clients should avoid attempting to claim more than the limit of PI insurance, equally professionals may seek to limit their financial liability by stating the maximum that can be claimed from them

is a multiple of the fees paid for the project or up to a certain fixed limit, which may or may not coincide with the limit of their (PI) insurance.

- **Payment.** Professionals may calculate their fees as a lump sum, as a percentage of construction cost, hourly rates or as a combination of these. Most professionals will require payment in stages, which may be monthly or when certain milestones have been achieved.

- **Termination.** Most professional services contracts continue to a successful conclusion but rarely there may be a need to terminate. The termination clause should specify the notice period to be given by each party and the consequences of termination.

- **Copyright.** Under the Copyright, Designs and Patents Act 1988, the copyright in a design rests with its author, that is the professional who produced it, unless the appointment document specifically vests copyright in the client. Even if this is not the case, provided the professional's fees have been paid according to the appointment, the client normally retains an implied licence to use the design for the purpose for which it was produced. This would not normally extend to the use of the design on another project without payment of a further licence fee.

- **Adjudication.** If a dispute arises on a project, it is sensible to have provisions already in place. Professional appointments for construction-related services are included under the Housing Grants, Construction and Regeneration Act 1996 and the Local Democracy, Economic Development and Construction Act 2009. They therefore contain provisions for dispute resolution by adjudication, which is a speedy and (supposedly) economic form of dispute resolution.

▶ **Dispute resolution.** Adjudication provides a temporarily binding method of dispute resolution, although it may become final and binding if the parties choose. Final and binding dispute resolution normally takes the form of litigation or arbitration. Arbitration has become expensive and long-winded but is completely private. Court proceedings are now much faster and more streamlined. Sometimes court proceedings can be cheaper than arbitration because the judge and venue are available at a low cost and the quality of judgments is very high. The choice between litigation and arbitration will now be a matter of how important confidentiality is to the parties.

Along with professionals, clients owe duties under a contract of appointment. The client will be under an obligation to confirm the priorities of their requirements under the brief and also to provide an indication of its budget and any time constraints. Clients should provide any information they have that could help the professionals, they are also obliged to take whatever actions and grant any approvals that allow the professional to discharge its services. The client is also obliged to pay for the services and this is likely to be in staged instalments. Excepting residential projects, the payment regime must comply with the Local Democracy, Economic Development and Construction Act 2009 with penalties for non-adherence.

Collateral warranties

Depending on the form of procurement, the client's appointed professionals may or may not provide services for the entire duration of the project. However, under design and build procurement the design team works for the client initially and then may be novated to the main contractor, from which point the professionals owe duties to the contractor rather than the client. The client or its funder may require any further design work done by the professionals to be covered under a collateral warranty. This means that although the professional is now working directly for the contractor, they may still owe the client a duty of care. Funders often require access to a professional's PI insurance in the event there are design defects in the building despite having no direct contractual link with them. The collateral warranty is the means that allows this to be done. It can also be used when a professional has worked directly for a developer client who wishes the first purchaser or tenant to be covered in the event of default, and again lenders may require this.

Construction contracts

Introduction

Effective clients should always enter into a formal construction contract with the constructor for any size of project. For a large project, lawyers may draft a bespoke contract but in most cases one of the standard forms, published by various bodies,[4] will suffice, possibly with some amendment.[5] If either party wishes to refer a dispute to adjudication, it is no longer necessary for a contract to be in writing but it is always advisable to have a written contract. A valid contract requires offer, acceptance, an intention to create legal relations, and consideration.

Clients choose from many different types of construction contract, depending on the size and complexity of the project, the selected procurement method and the degree of acceptable financial risk. It is beyond the scope of this chapter to review contract types in detail. Please refer to chapter 4, pages 32-53. Construction contracts have some features in common, including a set of general conditions applicable to any project and particular conditions that are project-specific. The powers of the contract administrator arise out of the contract conditions. Other matters such as subcontractors, insurances and dispute resolution will all normally be included.

Express and implied terms

The client's duties to provide information, allow access to the site and to make payments are usually expressed in the conditions. Most standard forms also make it obligatory to appoint a third party contract administrator. Having taken possession of the site, the contractor's progress should not be impeded or hindered (sometimes referred to as the prevention-principle), however it must take account of other contractors/suppliers for whom the employer is responsible.

Despite there being no duty that requires instructions, information, plans and drawings, etc to be issued in good time to the contractor, there is an implied obligation for it to cooperate. It is also implied that the employer shall not interfere with the functions or powers of his agents, nor exert influence when they perform their roles.

With regard to payment, there is an implied obligation to pay a reasonable sum, although this does not generally include an obligation to make payment by instalments unless the particular contract falls within the scope of the construction act where interim payments are a statutory right.

Over time the courts have implied many other obligations into construction contracts and generally the party seeking to rely upon an implied obligation must demonstrate reasonableness and necessity to do so. Any opposing views conversely need to be proven as well. Of course, certainty about contractual conditions can be achieved if they are expressly included in the contract, which is the case for most standard forms.

Payment

The Local Democracy Economic, Development and Construction Act 2009 imposes the requirement that construction contracts now have to provide the payment due date and a final date for payment. The client has to produce a payment notice and, if it wants to deduct money from any payment, produce what is known as a 'pay less' notice. There is a set timetable for the procedures after which the contractor may be entitled to take matters into their own hands. The final date for payment will normally be 14 days after the due date. Failure to operate the provisions correctly may entitle the contractor to the amounts it claims, whether or not the client considers they are due.

Insurance

The client should check that the contractor carries public liability insurance. The all-risks insurance is normally written in the joint names of the contractor and client and most forms of contract allow for the option of either to do so. For a new build it is normal for the contractor to take out the insurance, but the client may opt to do so. For works to existing buildings the client will normally have to take out insurance in the joint names of itself and the contractor. In the latter case the client should consult its insurer who will be able to recommend a suitable policy.

Practical completion (PC)

The first stage of the conclusion of a construction contract is practical completion, which is the stage at which the project is completed to the extent the client can take possession and begin using it. It does not necessarily mean the building is completely finished or free from defects, and even the courts have difficulty in defining PC. When the contract administrator awards PC, the contractor ceases to be responsible for insuring the works, and responsibility passes to the client or owner.

In large, complex projects or where a project is running late, a mechanism known as partial possession is used. This allows the client to take over and occupy part of the building whilst the rest is being completed.

The premature award of PC could deprive the client/employer of liquidated and ascertained damages, and incur the client additional cost due to the premature release of the retention monies. Practical completion is the precursor to the final inspection and the issue of the final certificate after the making good of any defects, after the expiry of the defects liability period that may be six or even 12 months after the award of PC.

In order to form an opinion on whether the alleged defective and outstanding works should prevent the issue of a certificate of practical completion, two recognised tests are considered of how PC should be defined that are in common use:

1 That the building must be complete, sufficient for the user to occupy and to use for its intended purpose, ie to enjoy 'beneficial occupation'. The existence of a limited number of minor defects would not prevent beneficial occupation and therefore would not prevent PC being achieved.

2 That the works must be complete, to the extent that there should be no patent defects other than defects that are 'trivial' or de minimis, or 'blemishes'.

In assessing the tests of practical completion, consideration should be given to:

▶ The nature and severity of the defect.

▶ Whether the defect might lead to further deterioration.

▶ The work required to remedy the defect.

▶ Whether such remedial work will disrupt the subsequent building user.

▶ Whether the health and safety (particularly in the event of fire) of the occupants is compromised by the defect and/or its remedial works.

Defective work

There are two types of defect: 'patent' and 'latent'. Patent defects are those apparent at a reasonable level of inspection, while latent defects are those which are concealed within the works and which may not become apparent for a number of years. Certainly latent defects should not be apparent prior to the issue of the final certificate. It is the contractor's responsibility to identify and rectify defects so that at the end of the works the architect is invited to award the certificate of practical completion with no defective work apparent. In practice, the architect will conduct regular inspections throughout the course of the work (usually preceding site meetings) noting any defective work and problems developing, so these may be addressed at the meeting and noted as action points. At the date of practical completion the work should be complete and free from patent defects, with only small 'blemishes' being allowed as the definition of de minimis defects.

The contractor must remedy all defects before the contract administrator can award practical completion. Only a relatively small number of *de minimis* defects are allowed to be present at PC, and these may be described as 'blemishes'. Following PC the period of defects liability starts and the contractor's liability to pay liquidated and ascertained damages (LADs) ends. The standard contractual period of defects liability is one year. The period must not be regarded by the contractor (and most certainly not by the contract administrator) as an opportunity to remedy defects that are visible prior to PC and to be remedied in the defects liability period which are outside the period in which the contactor is liable to pay LADs. In practice, however, PC may be certified subject to more obvious defects being attended to within an agreed timetable.

Letters of intent

Sometimes the need to make professional appointments, order products or proceed with construction is so urgent that the parties cannot wait for formal contracts to be created for signing. It may be that negotiations between the parties have yet to be concluded but activities have to start despite this. Such a letter declares the intention of the parties to enter into a formal contract and will normally contain details of what will happen should the parties not agree to enter into contract. Courts may recognise such a letter as a formal contract if it is not superseded by such a document. These letters should be followed as quickly as possible by fully detailed, formal contracts.

Termination

Construction contracts may be terminated either by the client or the contractor for a variety of reasons, including repudiatory conduct, insolvency, the prolonged suspension of the works or the contractor's corrupt behaviour. The client may wish to terminate his contract with the constructor because it failed to proceed regularly and diligently with the works. On termination, the client is entitled to take possession of the site and have the work completed by others. The contractor will prepare a final account and there will be no retention and payment obligations remaining under the contract will be allowed. Termination is a complex procedure and can be costly if not carried out properly. Clients in this position should consult a legal adviser before attempting to terminate.

Adjudication

Adjudication is a quick and relatively cheap way of resolving disputes. Any party to a construction contract has a statutory right to adjudication whether or not the contract made any provision for it. Parties to construction contracts have both a statutory and contractual right to adjudication as a means of dispute resolution. This does not apply to residential occupiers. If you wish to refer a dispute to adjudication, or if you find yourself on the receiving end of a referral, unless you are familiar with the process, it is best to immediately seek legal advice because the process takes place in a short time period. ■

Notes

[1] 92% of listed buildings are Grade II.

[2] 5.5% of listed buildings are in this category.

[3] Only 2.5% of listed buildings are in this category, which date from 2010.

[4] Contracts are available from the JCT, NEC, FIDIC, etc.

[5] Although frequently done, amending a standard form contract should be carried out with care.

Clients' troubles: avoidance and management

There is a common perception in society, promoted by an often hostile press searching for headlines, that lawyers are part of the problem – any problem. Keep away from them, and they will keep away from you. This is not mere taproom philosophy. Sir John Egan, in his 1998 report on the construction industry, 'Rethinking Construction', stated in terms that 'contracts (and, by implication, lawyers) can add significantly to the cost of a project and often add no value for the client'. He considered that should sufficient trust be established between employers and contractors, 'formal contract documents should gradually become obsolete'.

Back in the real world, that belief does few favours to either producers or consumers. Even where the relationship is good, the parties need to eliminate misunderstanding by clearly articulating their agreement as to what is to be done or delivered, at what cost, and within what period. But when relations are of the kind only too common in the commercial world – ie imbued with mutual suspicion and self-interest – early advice is crucial. Except in the simplest cases, contract formation has many pitfalls for the ignorant or unwary. It has been said that failing to crystalise and understand an intended contract is akin to putting your head in a guillotine and saying to your partner, the potential executioner, 'put it in the basket – I'll read it later'.

Construction contracts provide excellent examples of such pitfalls. Thus, standard forms cannot be relied upon for fairness. The RIBA Standard Agreement 2010 (2012 Revision) which, incidentally, excludes the client's entitlement to set-off for poor performance, is drafted with the primary intention of protecting architects from non-payment of their fees, not to balance risk between the parties. Again, in one case, the Court of Appeal said: 'indeed, if a prize were to be offered for a form of building contract which contained the most one-sided, obscurely and ineptly drafted clauses in the United Kingdom, the claim of this contract could hardly be ignored even if the RIBA Form of Contract was amongst the competitors' (the RIBA form referred to was the forerunner of the JCT standard form). Bespoke amendments to standard forms also have immense potential for confusion and error. For example, an uninformed employer might be tempted to remove employer's default as a reason for granting the contractor an extension of time, unaware that in doing so he may well be depriving himself of the right to liquidated damages (having potentially set time at large). And that is before one considers the need to ensure that construction contracts comply with the plethora of relevant regulations: Housing Grants, Construction and Regeneration Act 1996, Construction (Design and Management) Regulations 2015, Health and Safety at Work etc Act 1974, Public Procurement Regulations 2011, etc. Of course, many large employers and contractors are well served by their in-house expertise, but even they are likely to encounter complicated projects that require more specialist professional advice.

Furthermore, even the most carefully drafted contracts and the best commercial relationships cannot guarantee that disputes will not arise in the course of a project. Often, such disputes can be resolved without recourse to professionals, and sophisticated businessmen will be aware of the many methodologies available to facilitate settlement, such as mediation and the use of dispute review boards. Nevertheless, where agreement is, in the end, impossible, legal advice at an early stage will often pay dividends (by early stage it is meant the moment when the situation on-site starts to deteriorate, not when formal proceedings are imminent). The reasons for that are two-fold: substantive and presentational.

1 Substance: settlement is much more likely if a party has a clear understanding of its legal responsibilities and entitlements right from the beginning. The potential for unnecessary disaster is immense if it does not. A simple (but fatal)

error regularly seen in practice is a contractor's belief that a default by its employer (usually non-payment) is an excuse to leave site without notice. In fact, to abandon the project in such circumstances will often amount to a repudiatory breach of contract that may entitle the employer to substantial damages. However, a lawyer will be able to advise that, as an alternative, suspension of work (on notice) as permitted by the relevant legislation, may have the desired effect. Again, larger organisations may consider that they are sufficiently knowledgeable to avoid such errors, but the Law Reports are replete with cases involving major companies and public bodies litigating disputes that could probably have been avoided if competent advice had been sought at an early stage.

2 Presentation: there is a certain school of thought endemic in some organisations, from site operative to chairman, that laymen can draft effective correspondence as well as, or better than, any lawyer. The results can be disastrous. A recent example is the project architect whose draft letter was headed 'without prejudice and off the record', and went on to assert to the contractor, in the most aggressive terms, that the client was immensely wealthy and would pursue proceedings as a point of principle to the highest level, irrespective of cost. Since no offer was encompassed within the letter, the reference to its being 'without prejudice' was meaningless; and a simple claim to be writing 'off the record' is always ineffective. So this letter could have been shown to the court, potentially causing prejudice to the client's case as indicating an unreasonable attitude that, in the modern litigation world, could have had serious adverse costs consequences.

So, tempting as it may be to try to avoid or delay taking legal advice, the economic consequences of failure to do so can be as devastating as starting a building without investigating the ground conditions (albeit that the absence of legal advice does not usually lead to personal injury), and prudent organisations work closely with their legal teams so as to anticipate or minimise problems, recognising that leaving the law to chance is a false economy.

Chapter 12 How to keep out of trouble

by Murray Armes

"The effective client will aim to avoid disputes and will encourage the contractor to adopt the same approach. Dispute avoidance aims to tackle differences before they become crystallised disputes requiring a formal process of resolution."

Introduction

This book is all about the necessary techniques and habits of effective clients, and this particular chapter will show how a client can avoid disputes and, if that is not possible, next steps are described.

The need to avoid disputes

Why should clients want and need to avoid disputes? Primarily, they are costly in terms of money, time and reputation. An often hidden cost is the management time required to deal with disputes and the souring of relationships can make the going even harder. It is sometimes surprising just how early in the construction period disputes can arise and their negative impact may be felt throughout the project. It is worth taking into account that an average of 50% of all legal costs in the construction industry are related to disputes. In 10% of projects, 10% of total project costs are legal costs. Litigation will cost hundreds of thousands of pounds for a dispute that ends up in court, and an arbitration could cost at least that much.[1] This is all time and money that could be spent on the project itself or by improving margins and the financial performance of the industry.[2]

Effective clients are aware of the risks of moving too hastily through key stages, cost cutting particularly on fees or using the wrong criteria to make decisions. They are open to innovation – both in terms of products and new processes that have proven to be successful learning the lessons of others. They have the skills and experience to instruct their professional teams well and to evaluate their advice properly.

Their approach to projects both large and small is collaborative and their egos are suppressed. Effective clients think critically and sceptically and, above all, are sensible.

The nature and type of disputes

Disagreements and disputes arise out of uncertainty. However, few clients are prepared to acknowledge this inevitability at the outset under the influence of project optimism. This optimism usually flies in the face of the client's experiences in previous projects where confidence is seldom realised in practice. They will soon find out otherwise! Why are construction projects uncertain?

- Buildings are not mass-produced identical products – every project is a prototype. Even repeating the same design on a different site leads to uncertainty caused by unique site conditions.

- Project briefs may be inadequate or poorly defined.

- There are uncertainties in design because it is not possible to design everything before it is constructed (even using BIM) and the design is never complete until the building is finished. This is particularly appropriate for work to existing buildings.

- Choosing an ill-considered procurement process.

- Uncertainties in contractual arrangements and an imbalance of risk-sharing.

- Uncertainties in the construction phase including site conditions, weather, resources, and supply of materials, political and economic risks.

What type of disputes might arise?

▷ Disputes in construction typically arise from issues of:

- Contract interpretation.
- Quality of work.
- Progress.
- Information (quality of or lack of).
- Payment: ultimately (almost) all disputes arise as a result of financial issues.

▷ Construction projects involve teams of professionals who are connected only by the contractual arrangements specifically set out for the project. Disputes in construction commonly occur between:

- Employer and consultants.
- Employer and main contractor.
- Main contractor, subcontractors and suppliers.

What is a dispute?

Almost all contracts avoid defining a dispute, the exception being the FIDIC *Gold Book*[3] which defines a dispute as:

… *'Any situation where (a) one Party makes a claim against another Party; (b) the other Party rejects the claim in whole or in part; and (c) the first Party does not acquiesce, provided however that a failure by the other Party to oppose or respond to the claim, in whole or in part, may constitute a rejection if, in the circumstances, the DAB[4] or the arbitrator(s), as the case may be, deem it reasonable to do 'so'.*[5]

Disagreements are not always bad and may lead to innovation, but in construction projects clients should focus their efforts on ensuring that they do not end up in fully blown disputes. Different parties have different commercial interests: the client wants the best quality product at the best price and usually within the shortest time; however, the contractor sees the project as a commercial transaction from which it has to make as much profit as possible.[6] Quality will be a secondary consideration for most contractors, unless the provision of quality impacts the contractors' ability to be paid. The contractor will almost always want the programme to reflect the availability of resources and materials, and its need for cash flow. Recognising that the two main parties to a contract have different and sometimes opposing priorities means it is possible to understand how and why there are always two sides to a situation. Then there are the priorities of the various consultants.

Human issues

The human dimension is important at the project level; disputes are ultimately about people. Human traits leading to failures may be:

▷ Poor communications.

▷ An inability to communicate:

- Resorting to formal legal methods too quickly.
- Cultural differences and misunderstandings.
- Dysfunctional team (personality problems).

- Uncertainty making people defensive.
- Uncertainty leading to mistrust.
- Uncertainty preventing a proper assessment of risk.

The effective client will aim to avoid disputes and will encourage the contractor to adopt the same approach. Dispute avoidance aims to tackle differences before they become crystalised disputes requiring a formal process of resolution. Arguably, dispute avoidance is a state of mind but good project management is the basis for proactive dispute avoidance, that is a process with two main stages:

1 Management methods: focus on the construction process.

2 Non-escalation methods: focus on containing disputes.

Management methods

Management methods should be implemented throughout the project and should begin pre-contract, or as soon as possible thereafter. At this stage the only party involved will be the client and it is now that the client has an opportunity to set the tone for the whole project. One of the best ways of assessing the potential for disputes is to learn from previous projects. The next step is to be realistic about the possibility that problems will arise and consider what unpredictable scenarios might arise out of or during the course of the current project. This process is similar to the Risk Register included in NEC 3 Contracts[7] that allocates risk between the client (employer) and the contractor.

An effective client can help the project by providing all the information in their possession about the site and anything else that might adversely affect the project. It may be tempting simply to leave all the risk to the contractor but this will only lead to problems later. For instance, if you own the site, now is a good time to carry out a thorough site investigation and provide the information to the contractor.

Many clients want their project to be on-site as soon as possible and want to spend as little as possible on design. In Chapter 1, Figure 1.1 shows the relationship between design cost and the financial benefits of the project. The naïve client falls into this trap, often truncating the design period. Paying too little for design and giving the designers insufficient time to do a proper job may result in inadequate or inaccurate design information, and this almost always proves to be a fertile breeding ground for disputes.

Clients may choose a design/build form of procurement to reduce the time and cost of design. In principle, there is nothing wrong with this provided the pitfalls are recognised. The results of design/build can be capricious: some projects work out well, others do not. In particular, the quality of design, workmanship and materials can be unpredictable and sometimes disappointing. D&C procurement does not necessarily mean the project will be delivered quicker and because the contractor takes more risk, it is likely to prove more costly.

Designing everything before works start on-site will normally provide more certainty, however this approach may not suit the pace of a modern commercial environment. Whatever method of procurement is chosen, fewer disputes will arise when the production information, specifications and employer's requirements have been properly considered and well produced. In most cases it is worth investing a little more time and money in the project before the contract is let, to ensure that it is as well designed and coordinated as possible. The likelihood of disputes is reduced along with the need for changes during construction.

The choice of contract relates to the chosen of method of procurement. It is helpful to consider a contract that contains provisions for dispute avoidance as a precursor to formal dispute resolution procedures. The only standard form of construction contracts that make dispute avoidance mandatory are those published by FIDIC,3 which includes the *Gold Book*.[3] Their contracts are used for international construction projects and are less common in the UK. In the context of dispute avoidance, the NEC 3 contract was used for the London 2012 Olympics project.

Prudent management techniques should also be adopted at the second stage of the project where other parties, such as the main contractor, are involved. It is wise for clients to conduct due diligence on those they may be contracting with. Reputation, financial standing, organisation, people, methods and attitude are some of the topics to be assessed by clients.

It may be tempting for a client to adopt tender procedures that are not entirely fair to the potential contractor, for instance by not revealing factors that could affect the works and instead leaving the contractor to carry the risk. This will almost always result in disputes. The effective client will do its best to conduct a fair and open tender process, which is no guarantee that disputes will not arise.

All clients want best value for money and some also want the cheapest price. The effective client checks to ensure that the tender price is realistic. An unrealistically low price leaves the contractor with no alternative but to find ways to make a profit by making claims, and these usually result in disputes at best and at worst an insolvent contractor with all the attendant problems that entails.

Towards a collaborative, no-blame project culture

There is much to be said for encouraging a collaborative, no-blame regime. This is harder to establish than might at first appear because the parties each have different priorities, as described earlier in this chapter.

Such a collaborative, no-blame culture was successfully implemented on the Heathrow T5 project.[8] This was achieved by actively monitoring project costs and progress, but taking little account of any other alerts which may have resulted in delays to the work or prevented timely action being taken to keep the project on track. The key to active monitoring of the project is the recognition of problems as, and, if possible before they arise. Problems must be tackled as they arise and clients should not be tempted to leave dealing with them until the end of the contract, perhaps in the faint hope that they may go away.

Unfortunately, the sensible practice of making dispute avoidance a on-site process is rarely carried out in practice. Parties will leave problems until later, perhaps also fearful that airing problems might mean the deterioration of relationships. It is a natural human reaction to avoid dealing with conflict, but if the correct project management procedures are put in place they can be dealt with openly and promptly and not at the end of the project when the mistakes have already been made, positions have been taken, the seeds for disputes sown, and when memories have begun to fade.

Non-escalation methods

Having accepted that disagreements will arise, the next step is to contain them and ensure – if possible – that they do not progress to become formal disputes. Non-escalation is likely to take the form of a layered dispute resolution procedure beginning with negotiation. Roundtable negotiation is the cheapest, easiest and most direct method of dispute resolution that leaves the outcome entirely in the hands of the parties.

Solutions can be discussed and implemented that may reflect the terms of the contract or may depart entirely from those terms: it is entirely in the hands of the parties.

Reactive and proactive methods

It is not always possible for two or more parties to resolve issues without outside help in the form of a third party neutral. This assistance can take a number of forms and be more or less formal. Such assistance is preferable to the common mistake of implementing formal legal proceedings too quickly. Methods are:

- Mediation/ENE (reactive).
- Dispute Avoidance Panel (2012 Olympics, proactive).
- Ad hoc Statutory Adjudication (reactive).
- Ad hoc Dispute Board (reactive).
- Standing Dispute Board (proactive).
- Arbitration/litigation (reactive).
- Expert determination (reactive).

Reactive methods usually involve dispute resolution and are carried out after dispute has arisen. Mediation, which is known to be successful, sets out to be a non-confrontational form of dispute resolution, and is often preferable to court or arbitral proceeding because the parties retain some control over the outcome of their dispute.[9] Even at mediation the parties will have taken up positions and the process can become one of commercial (or personal) negotiation. The mediation process is without prejudice, unless the parties sign a settlement agreement, so creative methods of solving the dispute that might lie outside the contractual framework can be considered.

Adjudication and ad hoc dispute boards both involve adversarial processes and take place in a very short time. Although quick and relatively cheap, they can amount to 'rough justice'. However, neither process is final and binding, so if a party is sufficiently unhappy with the decision, it can proceed to litigation or arbitration, although most disputes end with the adjudication process. Remember that under most construction contracts the parties have a statutory (and usually also a contractual) right to adjudicate disputes. The process is quick,[10] so it is important to have dispute avoidance procedures in place long before any disputes are likely to arise.

One of the most effective methods of proactive dispute avoidance is the use of the standing dispute board.[11] The parties choose a panel of one or three experts and they start monitoring the project from the start of the contract. By making regular site visits and having meetings with both parties, the dispute board can proactively assist the parties in avoiding disputes, which at the request of the parties it can do by making informal recommendations or giving non-binding opinions. If a dispute does arise, the parties can refer it to the board for an adjudication decision.

Expert determination is a form of adjudication, which results in a final and binding decision, which is usually very hard to overturn. The process is not bound by the normal rules of natural justice, so while it can be very effective for some disputes the effective client should seek legal advice before implementing it.

Conclusion

There are many ways of getting into trouble because of the uncertainties that are inherent in the construction process. There are equally many ways of getting out of trouble, some more expensive in terms of time, money and reputations than others. However, the most effective way of keeping out of trouble is to implement proactive methods of dispute avoidance. If you do get into trouble, then consider all the options for getting out of it because there may be more effective (and private) ways than simply commencing arbitration or court proceedings. ∎

Notes

[1] Adjudication may be cheaper see pages 135-136.

[2] Construction industry profits are (currently) very low.

[3] FDIC is an organisation based in Geneva which, among other things, publishes international contracts, such as those included in the Gold Book.

[4] DAB: Dispute Adjudication Board see pages 135-136.

[5] FIDIC Sub Clause 1.1.31.

[6] The contractor may have other motivations, such as keeping their workforce active in lean times, but for the most part the priorities will be financial.

[7] NEC 3 Risk Register Clause 80.1.

[8] Heathrow Terminal 5 comprised 16 major projects and 140 sub-projects for a £4.3bn 270,000m2 main building, two satellite buildings and a new control tower. Team structure, effective communications, project culture and strong communications are seen as the criteria for delivering the project on time and to budget.

[9] Remember that once a dispute has been referred to a third party for a decision, it will be made on the basis of written and oral evidence, often without further reference to the parties. More information about dispute boards and their use can be obtained from the websites of the Dispute Resolution Board Foundation (DRBF) at www.drb.org, and the Chartered Institute of Arbitrators at www.ciarb.org.

[10] Adjudication takes place over 28 or 42 days.

[11] A hybrid standing dispute board was used for the London 2012 Olympics projects.

Perspective

by DDJ Stuart Kennedy

The construction client

The role of the client

The attitude of the client – or to use the contract term, 'the employer' – under a construction contract sets the tone for the whole project. An experienced and knowledgeable client has a much better understanding of the process of producing a new building and of what to expect from that process. For the less experienced or first time client, the role can be very stressful and frustrating, but it need not be so.

The starting point must be to properly identify and define what is required. This sounds obvious but so often difficulties arise when the finished product delivered by the contractor does not meet the client's expectations. There can be many reasons for this mismatch between what the client wants and what he gets, but failure to clearly define the specification from the outset is not uncommon. If the client is also to be the occupier of the building, thought must be given to the intended use of the building and the particular needs of the client, as well as the anticipated running and maintenance costs.

A key aspect for any client will be to appoint appropriate consultants, such as an architect, engineer and quantity surveyor. The number and type of consultants will depend upon the method of procurement of the building. Will it be a traditional contract with the design by the architect or perhaps design and construct by the contractor? Who will fulfil the role of project manager – the architect or a separate project manager? Ensure that the role of each consultant is defined in a written agreement and that there are no gaps between the scope of each role.

During the design process, the client should be actively involved and should be continuously reviewing the design to ensure it meets their needs.

Time and cost are fundamental issues. How long will the project take and how much will it cost? Both of these should be agreed at the outset and the continuously monitored for changes.

Using the correct contract form

The type of contract to be used should be carefully considered. The use of an inappropriate form of contract can create difficulties, as can bespoke amendments to standard forms. The choice of the form of contract will also be affected by the chosen method of procurement.

Knowing rights and obligations

When the contract has been chosen, and preferably before it is signed, the client should understand the terms of the contract and the rights and obligations of each party. Important terms cover issues such as payments, the completion date, extensions of time, liquidated damages, variations, quality and defects liability period. Understanding the contract can help avoid unreasonable expectations as to the performance of each party. An informed client should have realistic expectations as to what the contractor can and will supply under the contract. The client cannot re-write the contract later when it does not suit their changing needs. The contract is what it says – not what you wish it had said.

Be an informed client – get informed

Having employed all the right consultants, it is important to take and follow their advice. Be aware that an architect can have two separate and distinct roles, one as the designer of the works and the other as a certifier. In the latter role, the architect must act independently and not as the agent of the employer.

Avoid disputes

When the contract is produced and signed, the parties are the best of friends and hopefully they have no intention of getting into dispute. However, disputes are not uncommon in construction projects and so thought should be given as to how they can be avoided and, if they do arise, how they can be resolved.

Many disputes can be avoided if the parties behave in a reasonable manner and are realistic about their position and that of the other party. Recognising the potential for a dispute to escalate and the consequent cost in terms of time and money can help to focus on the desired outcome rather than the strict legal position.

When drafting the contract, think about what you will do if a dispute arises. Most construction contracts must, by law, incorporate a clause which entitles either party to refer a dispute to adjudication, which is a fast-track method of obtaining a binding but temporary decision. Many construction disputes are resolved by adjudication and do not go beyond that stage. However, in most instances a dispute can still be finally decided by arbitration or litigation, and clients should make a conscious decision about which to adopt. Other methods of resolving disputes include mediation, conciliation and expert determination. Some contracts (such as FIDIC) include an escalating dispute resolution process whereby disputes are first referred to a process of 'amicable settlement' (or negotiation), followed by mediation or conciliation and, finally, if the dispute has not been resolved at those two stages, to arbitration.

Chapter 13 The government client

by Janet Young

"The project must be underpinned with an economic options appraisal that assesses capital and revenue costs throughout the life of the project, risk and qualitative and quantitative benefits. The business case needs to assess a number of options for meeting the requirements."

Introduction

This chapter is for clients who have not worked in the government field before. It describes what is different about being a government client and signposts the reader to the available guidance.

Each central government department is responsible for managing its own estate with the Government Property Unit (GPU) in the Cabinet Office coordinating overall Government Estate Strategy and monitoring delivery. Each year the GPU publishes a 'State of the Estate' report that shows government departments' performance in the management of their estate and against sustainability targets. These two publications provide a good overview of the government's estate priorities and these and most other documents referred to in this chapter can be found on the gov.uk website.

Government construction and estate management

Government must lead by example, including commissioning and delivery of its own construction projects and how it manages its estate. The government's Construction 2025 strategy sets ambitious targets for cost and carbon reduction as well as reduced timelines for delivering projects. BIM is seen as crucial to achieving cost and time reductions and all government projects must achieve at least BIM Level 2 by 2016. The government is also keen to improve prompt payment to suppliers through project bank accounts, use collaborative procurement routes such as two-stage open book, and use integrated project insurance.

And with most property costs occurring during the operating life of the asset, Government Soft Landings is a policy that ensures BIM data is transferred into computer aided facility management (CAFM) systems to maintain the integrity and benefits of the building information dataset for the whole property life cycle.

The Ministry of Justice has embraced BIM for all its new build and major refurbishment projects, and increasingly for minor works projects. A dedicated BIM library with 98 standard component products has been set up that enables us to build any court or prison. New IT has been installed that will allow us to measure savings across the whole portfolio and, importantly, during the whole property life cycle, including FM.

The Greening Government Commitments targets require departments to reduce their carbon, waste and water use year on year. The construction strategy wants the UK to lead the world in low carbon and construction exports, and so new projects should actively assess potential for incorporating renewables and other low carbon products.

Characteristics of a government project

Government projects can be complex. Outcomes required could save lives, improve educational results or reduce crime, and there could be many different beneficiaries of projects. These beneficiaries can also be complex. A prison benefits society at large by supporting the rule of law but the location of new prisons also needs the support of the local community through the planning system. Furthermore, the local community may benefit from the prison, as it provides employment.

Government projects need to comply with relevant government policy objectives and demonstrate how others can achieve these objectives. So the project might include, for example, economic objectives like encouraging SMEs (small and medium-sized enterprises) to tender for elements of the project.

Project organisation

Parliament grants ministers the right to commit resources for their specific policy area and so a minister's role is to set the policy framework and priorities. It is the job of the civil service to deliver projects that meet these priorities. Both ministers and civil servants are accountable to the Public Accounts Committee (PAC) for how they spend money, including on projects.

For a project, the senior responsible officer (SRO) and the project director are key roles. The latter are responsible for the delivery of the project on time, to cost and of quality, and the SRO is accountable for the overall success of the project and ensuring that the benefits are delivered (lives saved, educational outcomes improved, crime reduced). Recent changes to civil service rules mean that SROs are now accountable directly to parliament.

Because of the complexity of government projects, it is essential to map out the governance that the project will need to navigate, as this could have a significant bearing on the overall schedule.

Getting approval

Getting consent to proceed with a government project requires a number of approvals. Projects must have a business case that complies with HM Treasury Green Book guidance, which requires five different cases. These are:

1 Strategic case for change: why do you need to do a project at all? What is the relative priority of the project for the organisation?

2 Economic case: does the project offer best value for the public purse? Which funding routes give the best NPV?

3 Financial case: is the project affordable within existing budgets?

4 Commercial case: is the proposed deal attractive to the market place? Can it be procured and is it commercially viable?

5 Management case: is the project achievable? Has the project team got the right skills and capacity to deliver it?

The project must be underpinned with an economic options appraisal that assesses capital and revenue costs throughout the life of the project, risk and qualitative and quantitative benefits. The business case needs to assess a number of options for meeting the requirements. In the case of a building project, this could include refurbishment, new build or minimal maintenance options. Different funding options will also need to be appraised, including both public sector and private sector funding and any hybrid options.

The preferred option will need to demonstrate that the project provides more benefits than other projects within the home department's portfolio – or across other government departments if capital budgets are being negotiated as part of the spending review. The question as to why the taxpayer should invest in this project as opposed to not investing at all or investing in other projects will require an answer.

If the project is considered to be a major project, defined on capital value/whole-life value terms, then the project will need an external assurance review from the Major Projects Authority and HM Treasury approval to the investment. If you are new to government projects, you are recommended to read the Major Project Approval and Assurance Guidance, published on gov.uk.

Public procurement rules

Public sector projects are by law subject to European Union procurement regulations and there is an increasing tendency for unsuccessful bidders to mount a legal challenge to procurement decisions with the potential for delay and additional cost. So it is essential to understand the broad requirements of the regulations and ensure that you have access to competent advice from experts with recent experience of public sector procurement.

The main impact on your project will be time: the regulations are there to ensure that contracts are awarded fairly and transparently, and that all potential bidders are treated equally. This takes time and there are no short cuts. The project will need to advertise its intentions to go to competition and must publish the results in the Official Journal of the EU (OJEU). The Crown Commercial Service is the government's agency that advises government departments about procurement and they publish considerable guidance on gov.uk

Almost all professional and constructor services are sourced via frameworks, which are competitively tendered in OJEU. That means that consultants and suppliers who want to work on government projects should ensure that they scan OJEU for notifications of forthcoming competitions.

Transparency and accountability

Government projects are funded from the public purse and overall responsibilities are described in HM Treasury's Managing Public Money available on gov.uk. The National Audit Office (NAO) is independent of government. It audits specific projects, as well as broader topics such as the effectiveness of asset management across the government estate. Their reports are submitted to the PAC and are published online. The government has committed to making as much public sector data as possible freely available. There is a rich source of data on www.data.gov.uk/data. The Freedom of Information Act is another source of information about government activity.

Conclusion

Many aspects of delivering a government project are the same. If you are new to government work, ensure that you familiarise yourself with the guidance, do not take shortcuts on procurement, include decision-making milestones in your project plan and make sure that you have access to public sector expertise in your team. ■

How to be an effective public sector client

The public and private sectors often regard each other as worlds apart. For many within the public sector, the private sector is seen as too focused on profit with little regard for the best interests of the public. Conversely, within the private sector there is often a view that the public sector is wasteful, with no concern for its need to make a fair profit in exchange for delivering services.

Both views are of course wrong but there are significant differences between the sectors, and the most effective public sector clients understand what these are, why they exist and how to balance the needs of each.

Above all else, private sector organisations must make a profit in order to survive. This imperative can sometimes cause them to fall into the trap of prioritising the bottom line to such an extent that good governance and due process is ignored, leading to the high-profile failures seen in the banking sector. It can also result in the delivery of poor quality services to customers. But the best private sector organisations do not make this mistake because they understand that long-term and sustainable success can only be achieved by maintaining effective governance, delivering value for money, building trust and providing high quality services to their clients rather than chasing short-term gain.

Public servants are accountable for every decision they take and are required to follow due process in their daily actions. However, at times these processes can cause delays, raise costs and at their worst create perverse outcomes caused by rigid adherence to the process, with no scope for the application of judgment by experienced managers trying to achieve the best possible outcomes. Recognising this issue is straightforward, however, addressing it is challenging. Any departure from what is considered acceptable due process within the public sector will be frowned upon by the civil service and the various bodies that hold them to account, and can lead to public censure, loss of employment and, in extreme cases, legal action against those involved.

The challenge to the public servant then is to maintain an appropriate balance between satisfying the requirements of due process while achieving results that recognise the need for

their private sector partners to return a fair and reasonable level of profit.

This can only be achieved with experience, constant effort, the full support of colleagues, senior managers and ministers with a clear and deliverable plan coupled with an effective communication strategy – all delivered by a fully resourced and excellent team.

None of this is a given and the root of many failed initiatives can be traced to gaps in one of more of these requirements.

Chapter 14 The client and the planning process

by Ruth Reed

"In ideal circumstances the project team develops the scheme design in collaboration with the authority and other stakeholders, and in consultation with the community."

Introduction

The British planning system is recognised throughout the world as an exemplar of democratic decision-making. However, because of the unpredictable nature of democracy coupled with diminishing resources, it can have a major impact on pre-construction programmes for large and small projects. An understanding of how delays can occur and the impact of different planning strategies on the programme will build greater certainty into the predicted timescales for the project. Nonetheless, the risk of delay cannot be eliminated completely.

The site

The likelihood of achieving planning consent can be greatly improved with careful preparation that can begin years before an application will be made. Sites can be put forward for allocation in the local plan for future use, enshrining their prospective development in planning policy.

For large, complex schemes, particularly those that require an estate appraisal, the alternative development strategies that are assessed as part of the preparation of the strategic brief will have different planning implications. For example, refurbishing existing buildings for alternative uses may have impacts outside the site boundary, such as intensification of traffic generation and parking requirements. Developing a greenfield site as an alternative may overcome some of these issues.

The planning balance

Once a site has been identified, the planning strategy should set out the planning balance. All planning decisions should be made in accordance with the development plan unless material considerations suggest otherwise. The development plan consists of the core strategy, local development framework and any neighbourhood plans or, where older plans are still in place, structure plans and local plans. As no application is entirely within policy, a balance is struck between those material considerations for the development and those against it, including any actual harm that might be caused.

Significant weight can be attached to economic, social and environmental considerations – the tenets of sustainability in national planning policy – so schemes that offer gains to the community on these grounds may overcome other objections. For example, providing employment opportunities may outweigh minor highways considerations. Each case is dealt with on its own merits and early consultation with the planning authority, if possible, will identify what can be offered to overcome objections.

The impact of project type and scale

An early consideration in the preparation of the strategic brief will be to decide whether it is necessary to establish a masterplan and bring forward the development in phases. This will add the time required to obtain outline planning consent to the project programme. However, it is possible to make a hybrid application with detailed consent sought for part of the site, which will enable the project to begin without compromising the possibility of refining the scheme in later phases as demand changes.

Major proposals carry particular requirements at planning. For example, they will require an environmental impact assessment and, if they have wider implications beyond the planning authority's boundaries, they may get called in by the secretary of state for determination. Take account of any permitted development rights.

Apart from Crown exemption, government – both local and national – carries no special exemption or priority within the planning system and applications should receive the same level of scrutiny as those made by the private sector.

For private applicants for schemes on public land, a planning performance agreement can be used to define the framework for a collaborative process between the LPA (local planning authority), the developer and its agents and key stakeholders such as the highways authority to develop, process and determine the application. Local authority planning departments are increasingly under-resourced and using PPAs (planning performance agreement) can focus the scarce resources on the project.

The pre-application process

The pre-application process should determine the validation requirements for the application and, as some can take many months to prepare, the sooner these are identified the less time will be lost validating the application. Environmental considerations, such as protected species like bats and reptiles, are seasonal users of land and buildings and the necessary data may not be available until a seasonal cycle is complete. Preparation of commercial viability appraisals may require months or even years of data collection to justify a change of use. A good strategic brief will have alerted the development team to the possible issues. They are then defined during pre-application discussions, following which the pre-application programme should be revised. Using consultants with local knowledge can help to predict the approach to constraints that an authority might take and contribute local knowledge of site conditions.

In ideal circumstances the project team develops the scheme design in collaboration with the authority and other stakeholders, and in consultation with the community.

Making an application

Some developers working at financial risk or within a two-stage tender process may be unwilling to appoint a full project team without the certainty of planning approval. In these circumstances they may make an application based on the RIBA Stage 2 Concept Design. This is risky because the full impacts of engineers' requirements on the design are not known and could result in material amendments to the planning approval and further delay once the developed design is complete. For design certainty, applications should be made with the Stage 3 Developed Design.

Although from 2016 all major public projects must use Level 2 BIM, the planning system is not yet capable of processing BIM models. Outputs in the conventional form of plans, sections and elevations are required. Applications are made through the planning portal, however many local authorities cannot receive payment electronically and this has to be submitted separately. Fees are payable on applications with exemptions and reductions in certain circumstances. For large schemes the fees can be as much as £250,000 and will be significant part of the scheme budget.

Additional costs to factor in are those sums due under the Community Infrastructure Levy, which if employed by the local authority are set out in pre-determined schedules. Other payments sought through legal agreements (Section 106 Agreements) are less predictable and should be defined through the pre-application process.

After submission: validation, refusal and appeal

Once the scheme is submitted there may be some delay due to operational issues within the authority before it is validated. However, the target determination date is calculated from the date of submission rather than validation. For major schemes this is 13 weeks. It is a target period, not a statutory requirement, and the authority's performance is monitored on its ability to meet this date. This can have several effects on authorities with limited resources. They may be under pressure to determine the application within the determination period and, if consultations are not complete, may refuse the application. They may request an extension to the period or they may let it run, prioritising schemes that they can determine within time.

For the agent and their client each presents a problem. For a refusal there will be reasons given that can be challenged at planning appeal or addressed in a revised submission. An extension of time may be symptomatic of a reduced level

of performance on the application; similarly non-determination can be frustrating as there are no concrete reasons why the scheme may not be acceptable. Frequently a combination of appeal for non-determination and a duplicate application can refocus the planning authority on the scheme. Alternatively, political pressure can be brought to bear on the department. For schemes with strong benefits to the economy and the community, a good case can be made to move swiftly to a resolution.

Planning permission conditions

Planning permission will come with conditions that will need to be discharged. Some before building work commences, some during construction, some before occupation and some in use. Scrutiny of these conditions is important because if any cannot be met it may result in funding agencies refusing to release funds or signing-off the project. If the conditions that will cause issues with completion do not meet one or more of the six tests, the condition should be challenged by applying to have it removed or, if that is unlikely to succeed, by appeal. All other conditions should be addressed promptly.

Managing change

One of the most significant challenges for any project is managing change within the design. Generally, significant design changes should be avoided, but for complex projects constructed over time in difficult market conditions for components and materials, this can be an impossible requirement to meet. Significant changes will require a complete reapplication for planning consent. ∎

The characteristics and skills of good and effective clients

HL Mencken said 100 years ago:

'For every complex problem there is an answer that is clear, simple, and wrong.'

And this is so true not only about what makes an effective client, but what makes a good design project. It is the little things done perfectly that fit into a cohesive whole and deliver the outcomes the specifier needed.

The good and effective client knows that clear, simple design is the result of a deep relationship between the client and designer, based on mutual trust and shared understanding of constraints, including budget, timescale, context and intended outcome.

The ideal client will have done his homework before the designer even comes on the scene.

He will be prepared to review similar design projects and talk with the clients for those projects about what worked, what did not work, and whether the designer was competent, outstanding or useless, and how you operate. So when you get the call, you hope the client knows what to expect from you.

The next step is to be able to articulate the vision for the project in terms that relate to the client's business. 'I want the resulting design to make my staff and my customers realise I value them, and increase our income but I only have £1m,' was the brief given to me for a trading room design project in London some years ago – the result was a radical (but low budget) design that changed the whole perception of customer contact in retail banking and generated a 100% plus uplift in business in the first year. Articulating a vision is not about second-guessing the design solution, it is about

Perspective

by Paull Robathan

creating a frame around the project into which the resulting picture fits neatly.

Once the designer knows what is needed in business terms, the work can start in earnest in collaboration with the client and his staff and customers. The process of coming to a design solution is itself a part of the end game because even before the design is realised, the level of respect and satisfaction with the project is being built up.

The perfect client engages just enough, but does not lead. The designer feels empowered to make recommendations that are not necessarily within the client's comfort zone, and should expect to have the opportunity to explain the rationale to a receptive and thoughtful audience.

Finally, when the work is done all I ask from a client is a simple 'thank you – that's just what I wanted,' followed by an open door to prospective future clients for a frank discussion about how hiring me worked out.

If all the above happens, then designer and client both walk away with heads held high, with a great new design that enhances both of their lives and gives satisfaction for a job well done.

Appendix I
Select Bibliography

The Aqua Group Guide to Procurement, Tendering & Contract Administration 2007, Blackwell Publishing (2007) and subsequent editions.

Architects Pocket Book, Architectural Press, 4th edition (2011)

Client Conversations: Insights into successful project outcomes, RIBA (2013)

The Institution of Engineering and Technology/ Centre for the Protection of National Infrastructure 'Resilience and Cyber Security of Technology in the Built Environment' (2013)

Berkovi, Jack, *Effective Client Management in Professional Services*, Gower (2014)

Bill, Peter, *Planet Property*, Matador (2013)

Blackmore, Courtenay, *The Client's Tale: The Role of the Client in Building Buildings*, RIBA Publications (1990), (OOP). Available at good second-hand bookshops

Bouchaghem, Dino, *Collaborative Working in Construction*, Routledge (2011)

Brett, Michael, *'Property and Money'*, Estates Gazette (2013)

Cresswell, H.B., *The Honeywood File & The Honeywood Settlement*, Butterworth Architecture (1990), (OOP). Available at good second-hand bookshops

Davies, Ian, *Construction Administration: RIBA Plan of Work Guide 2013*, RIBA Publications (2013)

Egan, Sir John, *Rethinking Construction: Report of the Construction Task Force*, London: HMSO (1998)

Evans, Huw, *Guide to the Building Regulations*, 2nd edition, RIBA Publishing (2014)

Eynon, John and CIOB, *The Design Manager's Handbook*, Wiley-Blackwell (2013)

Fairhead, Richard, *Information Exchanges: RIBA Plan of Work 2013 Guide*, RIBA Publishing (2015)

Gatti, Stefano, *Project Finance in Theory and Practice*, Academic Press (2012)

Harris, Frank, *Modern Construction Management*, 7th edition, Wiley-Blackwell (2013)

Havard, Dr Timothy, *Contemporary Property Development*, RIBA Publishing (2002)

Lupton, Sarah, *Cornes and Lupton's Design Liability in the Construction Industry*, 5th edition, Wiley-Blackwell (2013)

Reed, Richard and Sally Sims, *Property Development*, 6th edition, Routledge (2014)

Reed, Ruth, *Town Planning: RIBA Plan of Work Guide 2013*, RIBA Publishing (2014)

Sharratt, Fred and Peter Farrell, *Introduction to Construction Management*, Routledge (2015)

Sinclair, Dale, *Design Management: RIBA Plan of Work 2013*, RIBA Publishing (2014)

Sinclair, Dale, *Guide to using the RIBA Plan of Work 2013*, RIBA Publishing (2013)

Staiger, Roger, *Foundations of Real Estate Financial Modeling*, Routledge (2015)

Wevill, John, *Law in Practice: The RIBA Legal Handbook*, RIBA Publishing (2013)

Willars, Nick, *Project Leadership: RIBA Plan of Work Guide 2013*, RIBA Publishing (2014)

Alternative methods of dispute resolution in construction
www.designingbuildings.co.uk/wiki/Alternative_dispute_resolution_for_construction_ADR

Building Information Modelling (BIM) Task Group
www.bimtaskgroup.org

Construction Clients' Group Client Commitment Guides
ccg.constructingexcellence.org.uk/resources/publications/

Construction Clients' Group: Constructing Excellence
ccg.constructingexcellence.org.uk

Construction Industry Council
cic.org.uk/

Government Procurement Service
www.gov.uk/government/organisations/government-procurement-service

Health and Safety: The client's responsibilities
www.hse.gov.uk/construction/areyou/client.htm

Royal Institute of British Architects
www.architecture.com

The client's role in the sustainable building project
www.thenbs.com/topics/environment/articles/clientsRoleInTheBuildingProject.asp

Index

Note: page numbers in italics refer to figures;
page numbers in bold refer to tables.

Credits

Ben Hughes	42-43, 45, 47-49
Bennetts Associates	68, 76, 81
CABE (Creating Excellent Buildings: A Guide for Client, Eley et al 2003, CABE)	28
Constructing Excellence	4
Hufton+Crow	82
Joanne Eley	27
Morley von Sternberg	77
Peter Ullathorne	91
Susie Gray	11, 13, 18-19
Vodafone Limited	14
World Bank (Toolkit for Public-Private Partnerships in Roads & Highways, World Bank, Public Private Infrastructure Advisory Facility, March 2009)	51